Disclaimer

The publisher of this book is by no way associated with the National Institute of Standards and Technology (NIST). The NIST did not publish this book. It was published by 50 page publications under the public domain license.

50 Page Publications.

Book Title: Horizontal Convective Boiling of R134a, R1234yf/R134a, and R1234ze(E) within a Micro-Fin Tube with Extensive Measurement and Analysis Details

Book Author: Mark A. Kedzierski; Ki-Jung Park;

Book Abstract: This report presents local convective boiling measurements in a micro-fin tube for R134a and two low global warming potential (GWP) refrigerants: R1234yf/ R134a, 56/44 % mass and R1234ze(E). Water heating either in counterflow or in parallel flow with the test refrigerant was used to vary the heat flux for a given quality. The heat transfer coefficient of the three test fluids were compared at the same heat flux, saturated refrigerant temperature, and refrigerant mass flux using an existing correlation from the literature. The resulting comparison showed that refrigerant R134a exhibited the highest heat transfer performance in large part due to its higher thermal conductivity as compared to the tested low-GWP refrigerants. For the example case presented here, the heat transfer coefficient for R1234yf/ R134a (56/44) remains within 5 % of the heat transfer coefficient for R134a, having essentially identical performance for qualities less than 30 %. The heat transfer coefficient for R1234ze(E) is roughly 700 kW,K-1,m-2 (approximately 14 %) less than that of R134a for qualities greater than 30 %. The smaller heat transfer coefficient of R1234ze(E) as compared to that of R134a is primarily due to the 11 % smaller thermal conductivity and the 21 % smaller reduced pressure as compared to R134a at this test temperature. The measurements are important as part of the evaluation of low-GWP replacement refrigerants for R134a in unitary refrigeration and air-conditioning equipment.

Citation: NIST TN - 1807

Keywords: boiling, enhanced heat transfer, low-GWP, micro-fin, refrigerant mixtures

Horizontal Convective Boiling of R134a, R1234yf/R134a, and R1234ze(E) within a Micro-Fin Tube with Extensive Measurement and Analysis Details

Mark A. Kedzierski
Ki-Jung Park

**National Institute of
Standards and Technology**
U.S. Department of Commerce

Horizontal Convective Boiling of R134a, R1234yf/R134a, and R1234ze(E) within a Micro-Fin Tube with Extensive Measurement and Analysis Details

Mark A. Kedzierski
Ki-Jung Park
Energy and Environment Division
Engineering Laboratory

August 2013

U.S. Department of Commerce
Penny Pritzker, Secretary

National Institute of Standards and Technology
Patrick D. Gallagher, Under Secretary of Commerce for Standards and Technology and Director

National Institute of Standards and Technology Technical Note 1807
Natl. Inst. Stand. Technol. Tech. Note 1807, 54 pages (August 2013)
CODEN: NTNOEF

Horizontal Convective Boiling of R134a, R1234yf/R134a, and R1234ze(E) within a Micro-Fin Tube with Extensive Measurement and Analysis Details

M. A. Kedzierski
National Institute of Standards and Technology
Gaithersburg, MD 20899

K-J. Park
Korea Atomic Energy Research Institute
989-111 Daedeok-daero, Yuseong-gu, Daejeon, 305-353, Korea

ABSTRACT

This report presents local convective boiling measurements in a micro-fin tube for R134a and two low global warming potential (GWP) refrigerants: R1234yf/ R134a, 56/44 % mass and R1234ze(E). Water heating either in counterflow or in parallel flow with the test refrigerant was used to vary the heat flux for a given quality. The heat transfer coefficient of the three test fluids were compared at the same heat flux, saturated refrigerant temperature, and refrigerant mass flux using an existing correlation from the literature. The resulting comparison showed that refrigerant R134a exhibited the highest heat transfer performance in large part due to its higher thermal conductivity as compared to the tested low-GWP refrigerants. For the example case presented here, the heat transfer coefficient for R1234yf/ R134a (56/44) remains within 5 % of the heat transfer coefficient for R134a, having essentially identical performance for qualities less than 30 %. Similarly, the heat transfer coefficient for R1234ze(E) is essentially the same as that for R134a; however, it is roughly 700 $kW \cdot K^{-1} \cdot m^{-2}$ less than that of R134a for qualities less than 30 %. The smaller heat transfer coefficient of R1234ze(E) as compared to that of R134a is primarily due to the 11 % smaller thermal conductivity and the 21 % smaller reduced pressure as compared to R134a at this test temperature. The measurements are important as part of the evaluation of low-GWP replacement refrigerants for R134a in unitary refrigeration and air-conditioning equipment.

Keywords: boiling, enhanced heat transfer, low-GWP, micro-fin, refrigerant mixtures

TABLE OF CONTENTS

LIST OF TABLES

LIST OF FIGURES

INTRODUCTION

Internally enhanced tubes, like the micro-fin tube, are used by most manufacturers in the construction of evaporators and condensers for new unitary refrigeration and air-conditioning equipment. The reason for the micro-fin tube's hold on unitary equipment is that it provides the highest heat transfer with the lowest pressure drop of the commercially available internal enhancements (Webb and Kim, 2005). Most of the experimental measurements for evaporative heat transfer coefficients in the micro-fin tube have been done for traditional refrigerants like R134a. Pressure from the policies set by the Montreal Protocol (1987), the Kyoto Protocol (1997) and the European Mobile Directive (2006) have caused a recent shift to refrigerants with both zero ozone depletion potential (ODP) and low global warming potential (GWP). Johnson et al. (2012) reports that azeotropic R1234yf/ R134a (56/44) (i.e., XP10) [1] and R1234ze(E) are among the low-GWP refrigerants identified for evaluation by the Air-Conditioning, Heating, and Refrigeration Institute (AHRI) Low-GWP Alternative Refrigerants Evaluation Program as potential replacement refrigerants for R134a. The reason for this is that both R1234yf/ R134a (56/44) and R1234ze(E) have zero ODP and 100 year GWPs of approximately 600 and 6, respectively (Hickman, 2012 and Bitzer, 2012). Consequently, flow boiling heat transfer data for the micro-fin tube with R1234yf/ R134a (56/44) and R1234ze(E) are essential for the evaluation of their use for unitary applications.

Much of the relatively recent research on flow boiling in micro-fin tubes has been for traditional refrigerants. For example, Targanski and Cieslinski (2007) measured the evaporation heat transfer characteristics of R407C inside a micro-fin tube in the presence of oil. Zhang et al. (2007) measured the evaporation heat transfer coefficients of R417A and R22 inside a micro-fin tube and introduced a new heat transfer correlation to predict their values. Yun et al. (2002) examined existing experimental data and developed a model, which was validated for use with R22, R113, R123, R134a, and R410A and a variety of micro-fin tube geometries. Seo et al. (2000), Yu et al. (2002), and Kim et al. (2002) measured the flow boiling heat transfer coefficient in micro-fin tubes for R22, R134a, and R410A, respectively. Wellsandt and Vamling (2005) have investigated in-tube evaporation of R134a in a special type of micro-fin tube where the fin rifling, instead of being continuous, is arranged into V-grooves that resemble herringbones; hence, it is called the herringbone micro-fin tube. Oliver et al. (2004) have also studied the two-phase heat transfer performance of a herringbone and a standard 18-deg helical micro-fin tube with R22, R134a and R407C.

Because of the relatively recent introduction of R1234yf/ R134a (56/44) and R1234ze(E), measured heat transfer data in a micro-fin tube are not available in the literature for these refrigerants. The flow boiling measurements that presently exist for R1234ze(E), like those of Grauso et al. (2013) and Hossain et al. (2013), are for horizontal smooth tubes. One of the few recent works with R1234ze(E) in micro-fin tubes was by Koyama et al. (2011); however, this was a condensation study. Presently, there are no flow boiling measurements for R1234yf/ R134a (56/44) in a micro-fin tube. Consequently, the present study provides measured local flow boiling heat transfer for two low-GWP refrigerants (R1234yf/ R134a (56/44), and R1234ze(E)) and R134a in a micro-fin tube.

---

[1] Certain trade names and company products are mentioned in the text or identified in an illustration in order to adequately specify the experimental procedure and equipment used. In no case does such an identification imply recommendation or endorsement by the National Institute of Standards and Technology, nor does it imply that the products are necessarily the best available for the purpose.

EXPERIMENTAL APPARATUS

Figure 1 shows a sketch of the experimental apparatus used to establish and measure the convective boiling. The experimental test facility consisted of two main systems: the refrigerant loop and water loop. The refrigerant flow rate, pressure, and superheat were fixed at the inlet to the test section. The water flow rate and the inlet temperature were fixed to establish the overall refrigerant quality change in the test section. The water temperature drop, the tube wall temperature, the refrigerant temperatures, pressures, and pressure drops were measured at several axial locations along the test section. These measurements were used to calculate the local heat-transfer coefficient for the micro-fin tube.

The test section consisted of a pair of 3.34 m long, horizontal tubes connected by a U-bend. A fixed test pressure was maintained by balancing the refrigerant duty between the subcooler, the test section, and the evaporator. A magnetically coupled gear pump delivered the test refrigerant to the entrance of the test section with a few degrees of vapor superheat. Another magnetically coupled gear pump supplied a steady flow of water to the annulus of the test section. The inlet temperature of the water loop was held constant for each test with a water chilled heat exchanger and variable electric heaters. The refrigerant and water flow rates were controlled by varying the pump speeds using frequency inverters. Redundant flow rate measurements were made with Coriolis flowmeters and with turbine flowmeters for both the refrigerant and water sides.

Figure 2 shows a cross section of the test section with a detail of the micro-fin tube geometry. The test refrigerant flowed inside a micro-fin tube, while distilled water flowed either in parallel flow or counterflow to the refrigerant in the annulus that surrounded the micro-fin tube. Having some test in parallel flow and others in counterflow produced a broad range of heat fluxes at both low and high flow qualities. The annulus gap was 2.2 mm, and the micro-fin tube wall thickness was 0.3 mm. The micro fin tube had 60, 0.2 mm high fins with 18 degree helix angle. For this geometry, the cross sectional flow area was 60.8 mm^2 giving an equivalent smooth diameter (D_e) of 8.8 mm. The root diameter of the micro-fin tube was 8.91 mm. The inside-surface area per unit length of the tube was estimated to be 44.6 mm. The hydraulic diameter (D_h) was measured with a polar planimeter from a scaled drawing of the tube cross section and determined to be approximately 5.45 mm. The ratio of the inner surface area of the micro fin tube to the surface area of a smooth tube of the same D_e was 1.6. The fins rifled down the axis of the tube at a helix angle of 18° with respect to the tube axis.

Figure 3 provides a detailed description of the test section. The annulus was constructed by connecting a series of tubes with 14 pairs of stainless steel flanges. This construction permitted the measurement of both the outer micro-fin wall temperature and the water temperature drop as discussed in the following two paragraphs. The design also avoided abrupt discontinuities such as unheated portions of the test section and tube-wall "fins" between thermopile ends.

Figure 3 shows that thermocouple wires pass between 12 of the gasketed flange pairs to measure the refrigerant-tube wall temperature at ten locations on the top, side, and bottom of the tube wall. These locations were separated by 0.6 m on average, and they were located near the intersection of the shell flanges. In addition to these, thermocouples were also mounted next to the pressure taps near the middle of each test section length. The thermocouple junction was soldered to the outside surface and was sanded to a thickness of 0.5 mm. The leads were strapped to a thin non-electrically-conducting epoxy layer on the wall for a distance of 14.3 mm before they passed between a pair of the shell flanges. The wall temperature was corrected for a heat flux dependent fin effect. The correction was typically 0.05 K. Figure 3 also shows that a chain of thermopiles was used to measure the water temperature drop between each flange location. Each thermopile consisted of ten thermocouples in series, with the ten junctions at each end evenly spaced around the circumference of the annulus. Because the upstream junctions of one thermopile and the downstream junctions of another enter the annulus at the same axial location (except at the water inlet and outlet), the junctions of the adjacent piles were alternated around the circumference. A series of Teflon half-rings attached to the inner refrigerant tube centered the tube in the annulus. The half-rings were circumferentially baffled to mix the water flow. Mixing was further ensured by a high water Reynolds number (Kattan et al. 1995).

As shown in Fig. 3, six refrigerant pressure taps along the test section allowed the measurement of the upstream absolute pressure and five pressure drops along the test section. Two sets of two water pressure taps were used to measure the water pressure drop along each tube. Also, a sheathed thermocouple measured the refrigerant temperature at each end of the two refrigerant tubes, with the junction of each centered radially. Only the thermocouple at the inlet of the first tube was used in the calculations. The entire test section was wrapped with 5 cm of foam insulation to minimize heat transfer between the water and the ambient.

MEASUREMENTS

Table 1 shows the expanded measurement uncertainty (U) of the various measurements along with the range of each parameter in this study. The U was estimated with the law of propagation of uncertainty. All expanded measurement uncertainties are reported at the 95 % confidence level. The estimates shown in Table 1 are median values of U for the correlated data. Saturated refrigerant properties were evaluated at the measured saturation pressure with the REFPROP (Lemmon et al. 2010) equation of state, with the exception of the saturated temperature (T_s) and pressure (P_s) of R1234yf/ R134a (56/44), which was directly measured with a constant volume vessel, a temperature bath, a glass-rod standard platinum resistance thermometer (SPRT), and a pressure transducer. The measured temperature and pressure for R1234yf/ R134a (56/44) is presented in Table 2 and correlated below as:

$$T_s = \frac{1}{0.00571 - 3.25 \times 10^{-4} \ln P_s - 5.30 \times 10^{-6} \left(\ln P_s \right)^2} \tag{1}$$

The uncertainty in the temperature measurement was less than ± 0.01 K while the uncertainty in the pressure measurement was within ± 1 kPa.

The convective boiling heat transfer coefficient based on the actual inner surface area ($h_{2\phi}$) was calculated as:

$$h_{2\phi} = \frac{q''}{T_w - T_s} \tag{2}$$

where the measured wall temperatures (T_w) were fitted to their axial position to reduce the uncertainty in the measurement.

Figure 4 shows the estimated expanded uncertainty of the wall temperature fit for all the measurements as a function of thermodynamic quality. Figure 4 includes some data that was omitted from the correlation as explained in the Results Section. The uncertainty of roughly 90 % of the fitted wall temperatures was less than 0.5 K at the 95 % confidence level. The median of the uncertainty in T_w as shown in Table 1 was approximately 0.3 K.

The water temperature (T_f) was determined from the measured temperature change obtained from each thermopile and the inlet water temperature measurement. The water temperature gradient (dT_f/dz) was calculated with second-order finite difference equations using the measured water temperatures and their locations along the tube length z. The water temperature gradients were then fitted with a quadratic polynomial with respect to the tube length. As a check on the water temperature gradient calculation, Fig. 5 shows that the measured water temperatures (open circles) typically agreed with the integrated quadratic fit of the water temperature gradient (solid line) to within 0.2 K.

The fitted, local, axial water temperature gradient (dT_f/dz), the measured water mass flow rate ($\dot{m}_f$), and the properties of the water were used to calculate the local heat flux (q'') to the micro-fin tube based on the actual inner surface area:

$$q'' = \frac{\dot{m}_f}{p}\left(c_{p_f}\frac{dT_f}{dz} + v_f\frac{dP_f}{dz}\right) \tag{3}$$

where p is the wetted perimeter of the inside of the micro-fin tube. The specific heat (c_{pf}) and the specific volume (v_f) of the water were calculated locally as a function of the water temperature. The water pressure gradient (dP_f/dz) was linearly interpolated between the pressure taps to the location of the wall thermocouples. The pressure gradient term was typically less than 3 % of the temperature gradient term. Figure 6 plots the relative uncertainty of the heat flux measurement versus thermodynamic quality. As shown in Fig. 6, the uncertainty of the heat flux remains less than 3 % of the measured value, while the average uncertainty is approximately 1.5 % of the measured value.

Figure 7 shows example plots of the local heat flux as calculated from eq. (3) versus thermodynamic quality for both cases when the water and the refrigerant are in counterflow and parallel flow, respectively. Both heat flux profiles are for R134a at an all-liquid Reynolds number (Re) of roughly 7000 and a refrigerant reduced pressure of approximately 0.11. The discontinuity exhibited in the heat flux profiles is due to the change in refrigerant saturation temperature as caused by the adiabatic pressure drop in the bend that is used to transition from the first leg of the test section to the second leg. The decrease in the refrigerant saturation temperature causes an increase in the difference between the water and the refrigerant temperature, which leads to an increase in the local heat flux. For the counterflow case, the heat flux increases from approximately 3 kWm^{-2} at a quality near 0 to approximately 29 kWm^{-2} at a quality slightly greater than 0.8. The parallel flow case is nearly the mirror image of that for counterflow where the heat flux decreases from approximately 30 kWm^{-2} at a quality near 0.06 to approximately 3 kWm^{-2} at a quality slightly greater than 0.85.

The thermodynamic and transport properties were calculated with version 9.0 of REFPROP (Lemmon et al. 2010) while using enthalpy and pressure as inputs. The enthalpy of the refrigerant vapor at the inlet of the test section was calculated from its measured temperature and pressure. The subsequent drop in refrigerant enthalpy along the test section was calculated from the local heat flux and the measured refrigerant mass flow rate. The refrigerant pressures were measured at six pressure taps along the test section. The pressure was linearly interpolated between the taps. The average refrigerant temperature was varied between 1 °C and 3 °C with approximately 5 K of subcooling at the test section inlet. The open squares in Fig. 5 show the measured refrigerant temperature for an example test run with R1234yf/ R134a (56/44).

The local Nusselt number (Nu) was calculated using the hydraulic diameter and the heat transfer coefficient based on the actual inner surface area of the tube as:

$$\text{Nu} = \frac{h_{2\phi}D_h}{k_l} \tag{4}$$

Figure 8 shows the relative uncertainty of the Nu versus thermodynamic quality was between roughly 10 % and 25 %. Measurements of Nu with uncertainties greater than 25 % were discarded. The average uncertainty of Nu for presented data was approximately 18 % for all qualities.

RESULTS
The 451 data points generated in this study for R134a, R1234yf/ R134a (56/44), and R1234ze(E) are tabulated in Appendix A, which contains the Nusselt and all-liquid Reynolds numbers and other reduced data that are typically used to characterize flow boiling. Appendix B contains the raw data measurements including the heat flux and the wall and water temperatures and locations. The column entitled, "flow," provides a "C" or a "P" to indicate that the measurements were made for either counterflow or parallel flow between the refrigerant and water, respectively. All the parameters given in Appendix A and B are defined in the Nomenclature.

The measured local convective boiling Nusselt numbers (Nu) were compared to the pure-refrigerant (single component) version of the Hamilton et al. (2008) correlation:

$$\text{Nu} = 482.18\,\text{Re}^{0.3}\,\text{Pr}^{C_1}\left(\frac{P_s}{P_c}\right)^{C_2}\,\text{Bo}^{C_3}\,(-\log_{10}\frac{P_s}{P_c})^{C_4}\,M_w^{C_5} \tag{5}$$

where

$$C_1 = 0.51x_q$$
$$C_2 = 5.57x_q - 5.21x_q^2$$
$$C_3 = 0.54 - 1.56x_q + 1.42x_q^2$$
$$C_4 = -0.81 + 12.56x_q - 11.00x_q^2$$
$$C_5 = 0.25 - 0.035x_q^2$$

Here, the all-liquid Reynolds number (Re), the Boiling number (Bo), the liquid Prandtl number (Pr), the reduced pressure (P_s/P_c), and the quality (x_q) are all evaluated locally at the saturation temperature. The all-liquid Reynolds number and the Nusselt number are based on the hydraulic diameter (D_h). The Nusselt number is also based on the actual inner surface area of the tube.

The flow map of Yu et al. (2002) for micro-fin tubes was used to determine that approximately 87 % of the measurements were in annular or semi-annular flow. Manwell and Bergles (1990) suggest that the reason annular-like flow is a strong characteristic of micro-fin tubes is that the spiraling fins along the tube axis

5

encourage wetting of the upper tube wall.

Figure 9 shows a comparison between the boiling Nusselt numbers predicted with eq. (5) for the micro-fin tube to those measured here for R134a, R1234yf/ R134a (56/44) and R1234ze(E). Equation (5) predicts 77 % of the measured convective boiling Nusselt numbers for R134a, R1234yf/ R134a (56/44) and R1234ze in the micro-fin tube to within approximately ± 20 %. The measurements for each fluid are roughly centered about the mean of the correlation suggesting a lack of bias in the prediction due to the different fluids or some other cause.

Representative plots of the heat transfer coefficient ($h_{2\phi}$) versus thermodynamic quality (x_q) are given in Figs. 10 through 13. The solid lines are predictions for the present micro-fin tube geometry, which were obtained from the Hamilton et al. (2008) correlation given in eq. (5). The symbols are the measured data points, while the dashed lines provide the measurement uncertainty for a 95 % confidence level. The uncertainty in the heat transfer coefficient is shown to be roughly 1000 $WK^{-1} \cdot m^{-2}$ for most of data for qualities greater than 20 %. The uncertainty in the tube wall temperature is the greatest contributor to the uncertainty in the heat transfer coefficient.

Figure 10 shows the local heat transfer coefficient for R134a for Re = 6700 and $P_s/P_c = 0.09$ with counterflow between the refrigerant and the water. Half of the measurements are underpredicted by approximately 7.2 %, while the other half is overpredicted by approximately 6.2 %. Overall, the average difference between the measurements and the predictions is less that 1 %. The heat transfer coefficient increases with respect to quality, in large part, due to the increase of the local heat flux with respect to quality, which is a characteristic of counterflow.

Figure 11 shows the local heat transfer coefficient for R1234yf/ R134a (56/44) for Re = 5320 and $P_s/P_c = 0.11$ with counterflow between the refrigerant and the water. For qualities larger than 0.05 %, the measurements are predicted to within approximately 10 %. Overall, the average difference between the measurements and the predictions is less that 1 % for qualities larger than 0.05 %.

Figure 12 shows the local heat transfer coefficient for R1234ze(E) for Re = 9390 and $P_s/P_c = 0.08$ with counterflow between the refrigerant and the water. Seven of the measurements are overpredicted, on average, by approximately 7.7 %, while the remaining four measurements are underpredicted by approximately an average of 12.4 %. Overall, the average difference between the measurements and the predictions is less that 1 %.

Figure 13 shows the local heat transfer coefficient for R1234ze(E) for Re = 4570 and $P_s/P_c = 0.07$, which presents an example of the parallel flow condition. For qualities larger than 0.05 %, half of the measurements are underpredicted, on average, by approximately 9.2 %, while the other half are overpredicted by an average of approximately 7.9 %. Overall, the average difference between the measurements and the predictions is less that 1 %. For qualities less than 40 %, the heat transfer coefficient decreases with increasing quality. This is mainly caused by the decreasing heat flux with respect to quality, which is a characteristic of parallel flow.

Figure 14 uses the Hamilton et al. (2009) model to illustrate the relative heat transfer performance of R134a, R1234yf/ R134a (56/44), and R1234ze versus quality for the same saturated refrigerant temperature (T_s = 278 K), and the same refrigerant mass flux (G_r = 250 $kg \cdot m^{-2} \cdot s^{-1}$) for the present micro-fin tube geometry. Both counterflow and parallel flow conditions are shown. Counterflow is obtained by setting the heat flux to $q'' = 39x_q^{0.72}$ kW·m^{-2}, while parallel flow is obtained for $q'' = (31 - 32.6x_q)$ kW·m^{-2}. The heat flux profiles with respect to quality that were used to calculate the heat transfer coefficient are approximately equivalent to those shown in Fig. 7. Three different line styles for each flow condition are used to represent the predictions for the three different test fluids as labeled.

In general for counterflow, Fig. 14 shows that the boiling heat-transfer coefficient rapidly increases with increasing quality for qualities less than 20 %. For quality ranges between 20 % and 70 %, the rate of increase in the heat transfer coefficient with respect to increasing quality is roughly a fourth of that for qualities less than 20 %. For the example case presented here, the heat transfer coefficient for R1234yf/ R134a (56/44) remains with 5 % of the heat transfer coefficient for R134a, having essentially identical performance for qualities less than 30 %. For qualities greater than 30 %, the heat transfer coefficient for R1234ze(E) is roughly 700 kW·K^{-1}·m^{-2} less than that of R134a. The smaller heat transfer coefficient of R1234ze(E) as compared to that of R134a is primarily due to the 11 % smaller thermal conductivity and the 21 % smaller reduced pressure as compared to R134a at this test temperature. The favorable performance of R134a as compared to the tow low-GWP refrigerants examined here is primarily due to the larger liquid thermal conductivity.

For parallel flow, Fig. 14 shows nearly the same relative and absolute performance for qualities greater than 20 %. However, the influence of the larger heat flux is evident for qualities less than 20 % for the parallel flow condition. For qualities less than 20 %, it is likely that nucleate boiling may more influential in determining the magnitude of the heat transfer coefficient than it is for the counterflow condition. In this region, the heat transfer coefficient is shown to decrease with increasing quality as the nucleate boiling becomes suppressed with the growing presence of annular flow. Otherwise, the heat transfer coefficient for parallel flow is rather constant with respect to quality varying no more that ± 11 % from its mean value over the illustrated quality range.

CONCLUSIONS
Local convective boiling heat transfer measurements for two low-GWP refrigerants and R134a in a fluid heated micro-fin tube were presented. The measured convective boiling Nusselt numbers for all of the test refrigerants were compared to an existing correlation from the literature. Approximately, 77 % of the measurements were predicted to within ± 20 % and centered about the mean prediction.

In general, the measured boiling heat-transfer coefficient increased with increasing qualities for counterflow between the refrigerant and the water. In contrast, for parallel flow, the measured heat transfer coefficient was relatively constant. The heat transfer coefficient of the three test fluids were compared at the same heat flux, saturated refrigerant temperature, and refrigerant mass flux by using the correlation from the literature that was validated with the measurements. The resulting comparison showed that refrigerant R134a exhibited the highest heat transfer performance in large part due to its higher thermal conductivity as compared to the tested low-GWP refrigerants. For the example case presented here, the heat transfer coefficient for R1234yf/ R134a (56/44) remains within 5 % of the heat transfer coefficient for R134a, having essentially identical performance for qualities less than 30 %. Similarly, the heat transfer coefficient for R1234ze(E) is essentially the same as that for R134a; however, it is roughly 700 kWK^{-1}m^{-2} less than that of R134a for qualities less than 30 %. The smaller heat transfer coefficient of R1234ze as compared to that of R134a is primarily due to the 11 % smaller thermal conductivity and the 21 % smaller reduced pressure as compared to R134a at this test temperature.

ACKNOWLEDGEMENTS

This work was funded by NIST. The authors thank the following for their constructive criticism of the first draft of the manuscript: Dr. P. Domanski, and Dr. A. Persily from NIST, and Dr. D. Han from LACC-JNK Inc. Thanks also go to Wolverine Tube, Inc., for supplying the Turbo-A, micro-fin tube for the test section, and to Dupont and Honeywell for supply the R1234yf/ R134a (56/44), and R1234ze(E) test refrigerants, respectively.

NOMENCLATURE
<u>English symbols</u>

A_c cross-sectional area

Bo   local boiling number, $\dfrac{q''}{G_r i_{fg}}$

c_p specific heat (J/kg·K)
C coefficients given in eq. (5)

D_e equivalent inner diameter of smooth tube, $\sqrt{\dfrac{4 A_c}{\pi}}$ (m)

D_h hydraulic diameter of micro-fin tube (m)
e fin height (mm)
G total mass velocity (kg/m²·s)
$h_{2\phi}$ local two-phase heat-transfer coefficient (W/m²·K)
i_{fg} latent heat of vaporization (J/kg)
k refrigerant thermal conductivity (W/m·K)
Nu   local Nusselt number based on D_h
$\dot{m}$ mass flow rate (kg/s)
M_w molar mass (g/mole)
p wetted perimeter (m)
P local fluid pressure (Pa)

Pr   liquid refrigerant Prandtl number $\left.\dfrac{c_p \mu}{k}\right|_{r,l}$

q'' local heat flux based on A_i (W/m²)

Re   all liquid, refrigerant Reynolds number based on $D_h = \dfrac{G_r D_h}{\mu_{r,l}}$

S_v non-dimensional refrigerant specific volume given in Appendix: $\dfrac{v_v - v_l}{v}$

s distance between fins (mm)
T temperature (K)
t_b bottom thickness of fin (mm)
t_w tube wall thickness (mm)
U expanded relative uncertainty
x_q thermodynamic mass quality
z axial distance (m)

<u>Greek symbols</u>

α helix angle (°)

β fin angle (°)

ΔT_s T_s - T_w (K)
μ viscosity (Pa·s)
v specific volume, $x_q v_v + (1 - x_q) v_l$ (m³/kg)

Subscripts

c	critical condition
f	water
l	liquid
p	prediction
r	refrigerant
s	saturated state
v	vapor
w	heat transfer surface

REFERENCES
Bitzer Kuhlmaschinenbau GmbH, 2012, "Refrigerant report 17"

European Mobile Directive, 2006, "Directive 2006/40/EC of The European Parliament & of the Council of 17 May 2006 Relating to Emissions from Air-Conditioning Systems in Motor Vehicles & Amending Council Directive 70/156/EC," <u>Official Journal of the European Union</u>, Vol. 49, No. L 161, pp.12-18.

Grauso, S., Mastrullo, R., Mauro, A. W., Thome, J.R., and Vanoli, G. P., 2013, "Flow Pattern Map, Heat Transfer and Pressure Drops During Evaporation of R1234ze(E) and R134a in a Horizontal, Circular Smooth Tube: Experiments and Assessment of Predictive Methods," <u>International Journal of Refrigeration</u>, Vol. 36, pp. 478-491.

Hamilton, L. J., Kedzierski, M. A, and Kaul, M. P., 2008, "Horizontal Convective Boiling of Pure and Mixed Refrigerants within a Micro-Fin Tube," <u>Journal of Enhanced Heat Transfer</u>, Vol. 15, No. 3, pp. 211-226.

Hickman, K. E., 2012, "Alternatives to High GWP HFC Refrigerants: Chiller Applications," *Proceedings of ASHRAE/NIST Refrigerants Conference*, Gaithersburg, MD, USA.

Hossain, Md. A, Onaka, Y., Afroz, H., M.M., and Miyara, A., 2013, "Heat Transfer During Evaporation of R1234ze(E), R32, R410A, and a Mixture of R1234ze(E) and R32 Inside a Horizontal Smooth Tube," <u>International Journal of Refrigeration</u>, Vol. 36, pp. 465-477.

Johnson, P. A., Wang, X., Amrane, K., 2012, "AHRI Low-GWP Alternative Refrigerant Evaluation Program," *Proceedings of ASHRAE/NIST Refrigerants Conference*, Gaithersburg, MD, USA.

Kattan, N., Favret, D., and Thome, J. R., 1995, "R-502 and Two Near-Azeotropic Alternatives: Part I - In Tube Flow-Boiling Tests," <u>ASHRAE Trans.</u>, Vol. 101, Pt. 1, pp. 491-508.

Kim, Y., Seo, K., and Chung, J., 2002, "Evaporation Heat Transfer Characteristics of R-410A in 7 and 9.52 mm Smooth/Micro-Fin Tubes," <u>International Journal of Refrigeration</u>, Vol. 25, pp. 716-730.

Koyama, S., Baba, D., and Nakahata, H., 2011, "Experimental Study on Heat Transfer and Pressure Drop Characteristics of Pure Refrigerant R1234ze(E) Condensing in a Horizontal Micro-Fin Tube," *International Congress of Refrigeration, Beijing Prague, Czech Republic*, ID: 306.

Kyoto Protocol, 1997, "United Nations Framework Convention on Climate Change," United Nations (UN), New York, NY, USA.

Lemmon, E. W., Huber, M. L., and McLinden, M. O., 2010, NIST Standard Reference Database 23, Version 9.0. Private Communications with McLinden, National Institute of Standards and Technology, Boulder, CO.

Manwell, S.P., and Bergles, A.E., 1990, "Gas-liquid flow patterns in refrigerant-oil mixtures, <u>ASHRAE Transactions</u>, Vol. 96, Part 2, p. 456-464.

Montreal Protocol, 1987, "Montreal Protocol on Substances that Deplete the Ozone Layer," United Nations (UN), New York, NY, USA (1987 with subsequent amendments).

Olivier, J. A., Liebenberg, L., Kedzierski, M. A., and Meyer, J. P., 2004, "Pressure Drop During Refrigerant Condensation Inside Horizontal Smooth, Helical Micro-Fin, and Herringbone Micro-Fin Tubes," Journal of Heat Transfer, Vol. 126, pp. 687-696.

Seo, K., and Kim, Y., 2000, "Evaporation Heat Transfer and Pressure Drop of R-22 in 7 and 9.52 mm Smooth/Micro-Fin Tubes," International Journal of Heat and Mass Transfer, Vol. 43, pp. 2869-2882.

Targanski, W., Cieslinski, J. T., 2007, "Evaporation of R407C/Oil Mixtures Inside Corrugated and Micro-Fin Tubes" Applied Thermal Engineering, Vol. 27, No. 13, pp. 2226–2232.

Webb, R. L., and Kim, N-H., 2005, Principles of Enhanced Heat Transfer, 2nd ed., Taylor & Francis, New York.

Wellsandt, S., and Vamling, L. 2005, "Evaporation of R134a in a Horizontal Herringbone Microfin Tube: Heat Transfer and Pressure Drop," International Journal of Refrigeration, Vol. 28, pp. 889–900.

Yu, M., Lin, T., and Tseng, C., 2002, Heat Transfer and Flow Pattern During Two-phase Flow Boiling of R-134a in Horizontal Smooth and Micro-fin Tubes, International Journal of Refrigeration, Vol. 25, pp. 789-798.

Yun, R., Kim, Y., Seo, K., and Kim, H., 2002, "A generalized correlation for evaporation heat transfer of refrigerants in micro-fin tubes," International Journal of Heat and Mass Transfer, Vol. 45, pp. 2003-2010.

Zhang, X., Zhang, X., and Yuan, X., 2007, "Evaporation Heat Transfer Coefficients of R417a in Horizontal Smooth and Microfin Tubes," *International Congress of Refrigeration, Beijing*, ICR07-B1-228.

Table 1 Median estimated 95 % relative expanded uncertainties for measurements (U)

Parameter	Minimum	Maximum	$U \%$
G_r [kg/m$^2 \cdot$s]	100	418	2.0
T_s [K]	293.0	323.0	0.1 (0.3 K)
P [kPa]	270	450	1.5
T_w [K]	279.0	293.0	0.1 (0.25 K)
$\dot{m}_f$ [kg/s]	0.010	0.030	2.0
T_f [K]	281.0	321.0	0.1
P_f [kPa]	200	110	1.0
q'' [kW/m^2]	2.6	42.2	5.1
dT_f/dz [K/m]	0.016	0.43	5.2
Nu	112	460	16.4
Re	2191	10800	4.0
Bo	0.000037	0.00063	16.0
Pr	3.6	4.2	2.0
P_s/P_c	0.06	0.12	2.0
x_q	0.003	0.82	8.0
ΔT_s [K]	1.3	7.6	15.2 (0.44 K)

Table 2 Measured saturated temperature and saturated pressure of R1234yf/ R134a (56/44)

T_s (K)	P_s (kPa)	T_s (K)	P_s (kPa)	T_s (K)	P_s (kPa)	T_s (K)	P_s (kPa)	T_s (K)	P_s (kPa)
303.05	809.3	284.44	467.8	274.76	339.9	283.74	457.5	290.03	556.4
302.55	798.8	283.93	460.4	275.08	343.7	283.73	457.5	290.33	561.4
302.06	788.5	283.44	453.2	275.41	347.5	283.74	457.5	290.62	566.4
301.56	778.0	282.94	446.0	275.73	351.4	283.74	457.5	290.91	571.4
301.06	767.6	282.44	438.9	276.05	355.2	283.74	457.5	291.20	576.4
300.58	757.1	281.93	431.8	276.37	359.0	283.73	457.5	291.50	581.5
300.09	746.6	281.43	424.8	276.69	362.9	283.74	457.5	291.79	586.7
299.59	736.3	280.92	417.9	277.01	366.8	283.73	457.5	292.09	591.8
299.10	726.2	280.42	411.1	277.32	370.8	283.74	457.5	292.37	597.0
298.61	716.1	279.90	404.1	277.63	374.7	283.73	457.5	292.66	602.2
298.12	706.2	279.39	397.4	277.94	378.5	283.73	457.5	292.96	607.5
297.63	696.6	278.88	390.8	278.25	382.6	283.73	457.6	293.25	612.8
297.14	686.8	278.36	384.0	278.57	386.6	283.74	457.6	293.54	618.1
296.66	677.2	277.84	377.3	278.89	390.6	283.73	457.5	293.83	623.4
296.18	667.9	277.82	377.1	279.19	394.6	283.73	457.5	294.13	628.9
295.68	658.3	277.14	368.7	279.50	398.6	283.73	457.6	294.42	634.3
295.19	649.0	276.26	358.0	279.81	402.6	283.73	457.6	294.71	639.8
294.71	639.9	275.73	351.6	280.11	406.8	283.83	459.0	295.00	645.2
294.22	630.7	275.19	345.2	280.42	410.9	284.14	463.3	295.29	650.7
293.73	621.9	274.65	338.9	280.73	415.0	284.44	467.8	295.58	656.3
293.25	613.1	274.11	332.6	281.02	419.0	284.74	472.2	295.88	662.0
292.76	604.2	273.55	326.4	281.33	423.3	285.03	476.6	296.18	667.8
292.27	595.6	273.00	320.2	281.64	427.5	285.33	481.0	296.47	673.4
291.79	587.0	272.45	314.1	281.94	431.7	285.62	485.5	296.76	679.0
291.29	578.3	271.88	307.8	282.24	435.9	285.92	490.1	297.05	684.8
290.81	570.0	271.31	301.8	282.54	440.1	286.22	494.7	297.35	690.7
290.33	561.7	270.74	295.8	282.84	444.4	286.51	499.2	297.64	696.6
289.83	553.3	271.13	299.6	283.14	448.8	286.80	503.8	297.93	702.5
289.35	545.1	271.15	299.8	283.44	453.1	287.10	508.4	298.22	708.3
288.86	537.1	271.44	302.9	283.74	457.4	287.39	513.1	298.52	714.3
288.37	529.0	271.76	306.4	283.74	457.4	287.69	517.8	298.81	720.3
287.88	521.1	272.11	310.1	283.74	457.4	287.98	522.5	299.11	726.4
287.39	513.3	272.45	313.7	283.74	457.5	288.27	527.2	299.40	732.4
286.90	505.5	272.79	317.5	283.74	457.5	288.57	532.0	299.69	738.6
286.41	497.9	273.12	321.2	283.74	457.5	288.86	536.9	299.99	744.4
285.92	490.3	273.45	324.9	283.74	457.5	289.16	541.7	300.29	750.3
285.42	482.6	273.78	328.7	283.74	457.5	289.45	546.5	300.58	756.5
284.93	475.3	274.11	332.4	283.74	457.5	289.74	551.4	300.87	763.0
						301.47	775.8	301.17	769.2

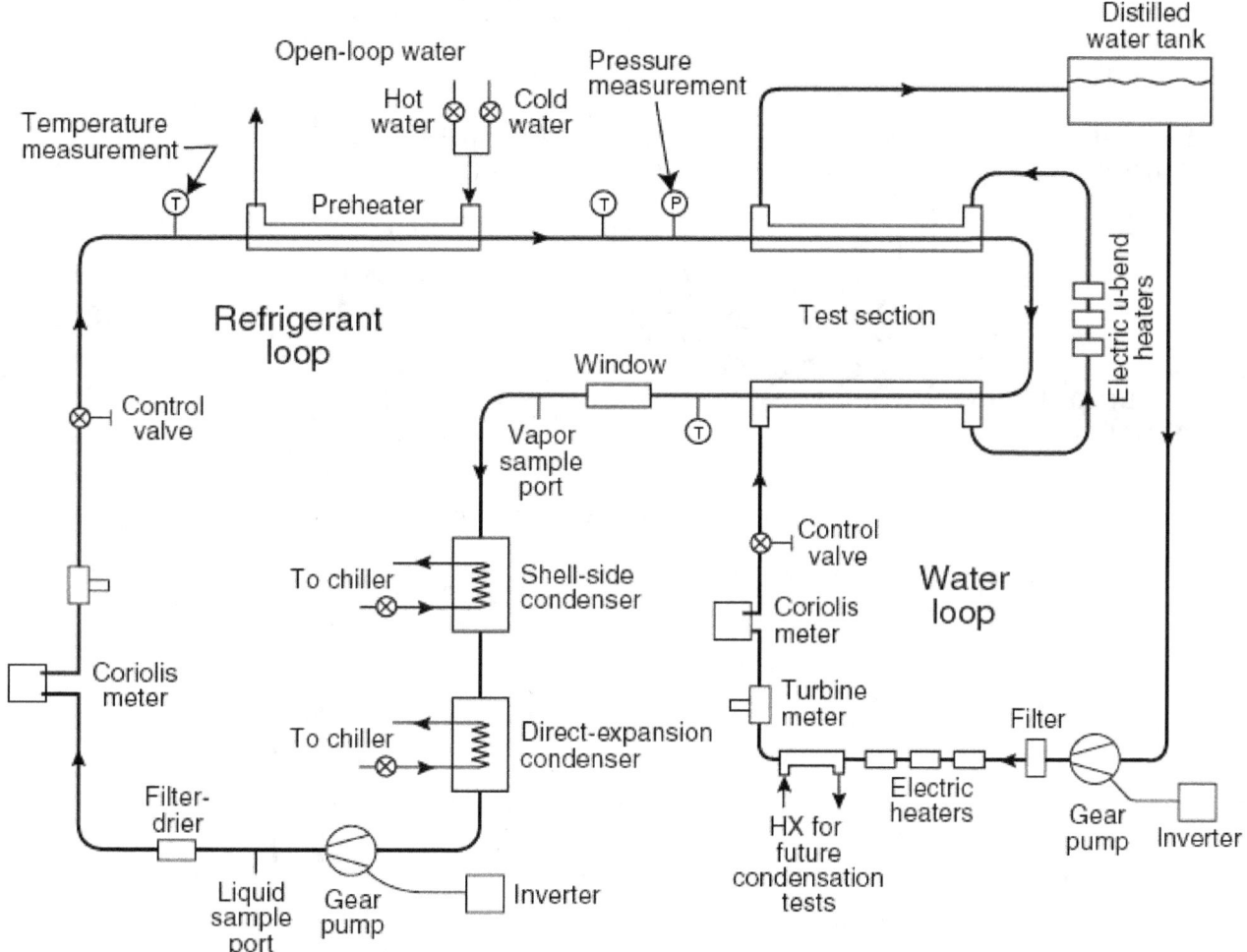

Figure 1 Schematic of test rig

15

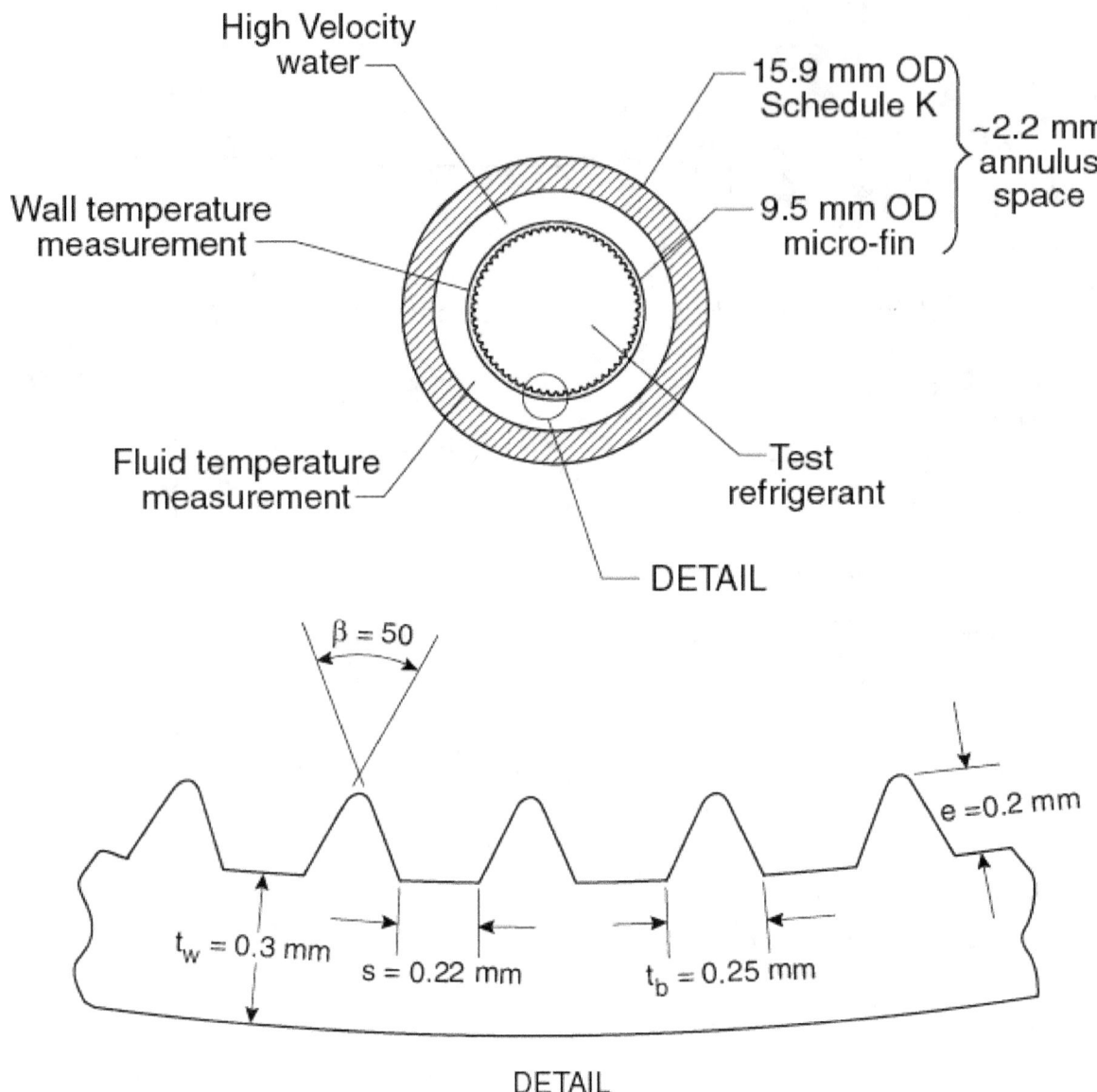

Figure 2 Test section cross section

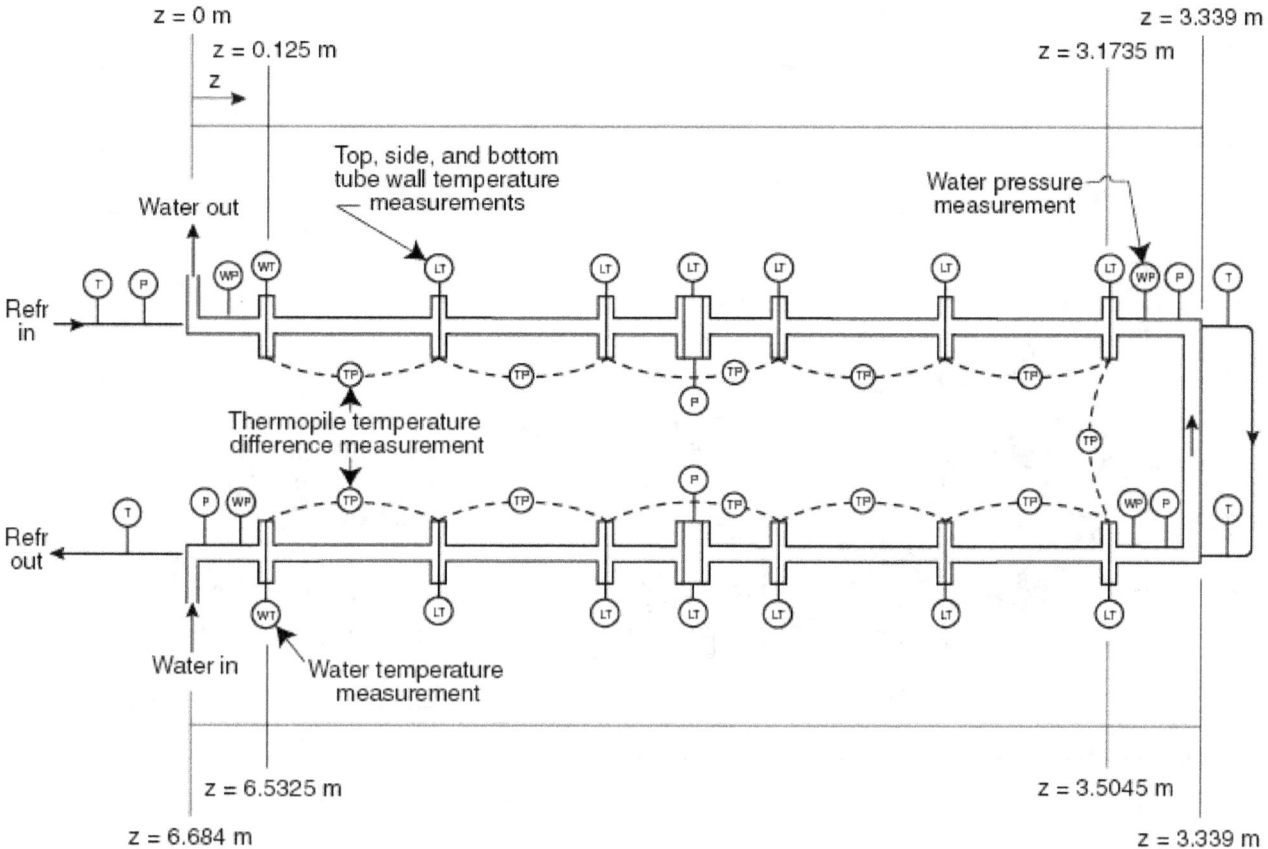

Figure 3 Detailed schematic of test section

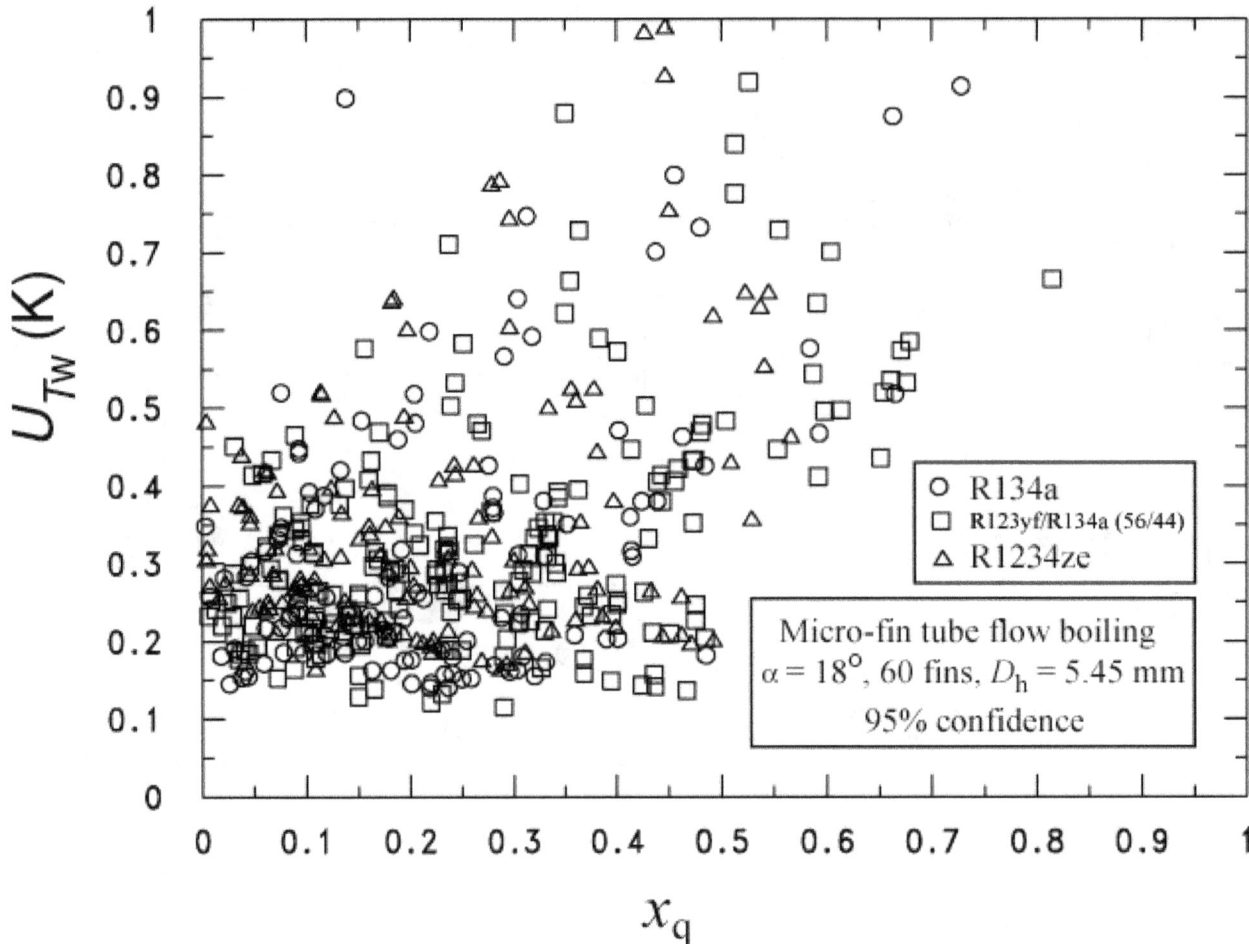

Figure 4 Relative uncertainty of inner wall temperature

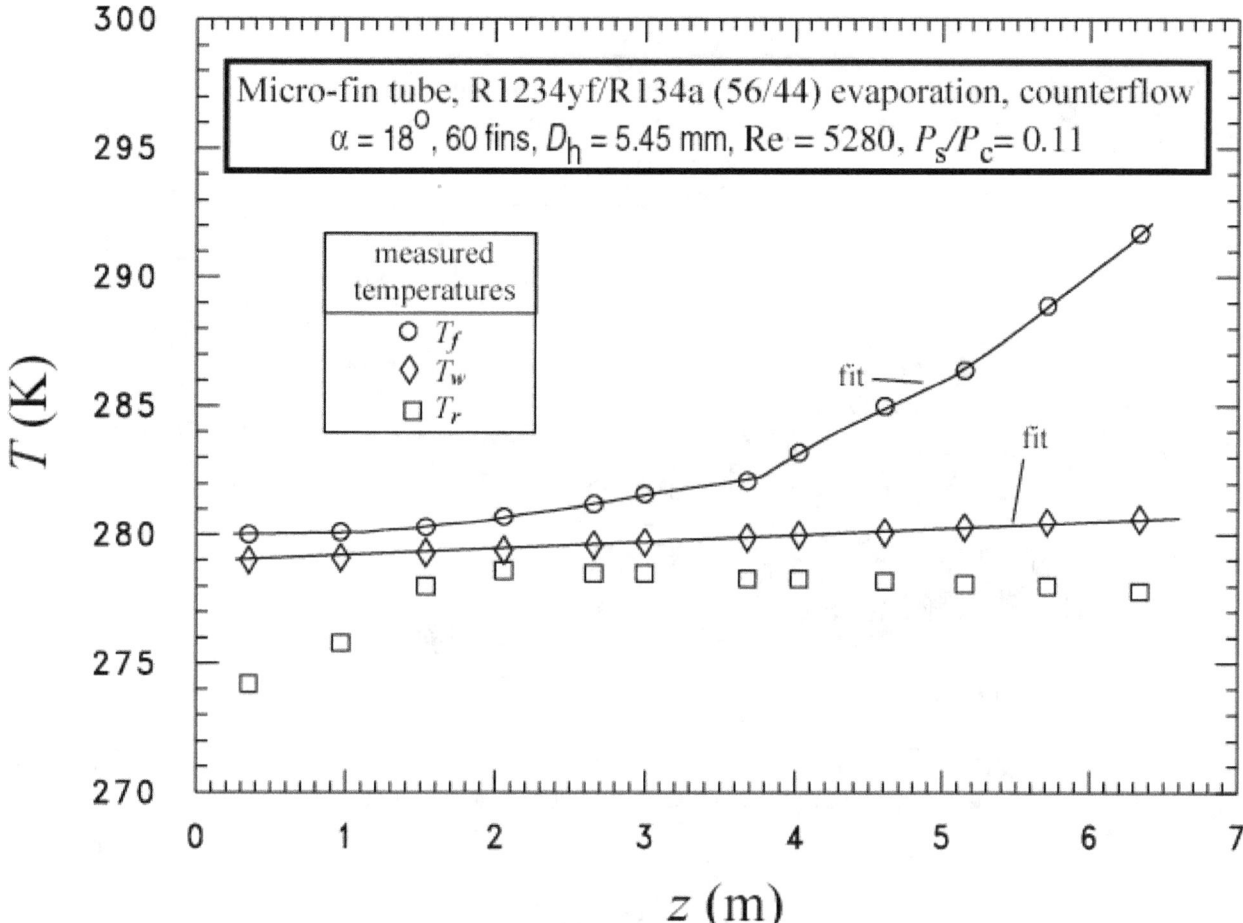

Figure 5 Counterflow temperature profiles for a R1234yf/ R134a (56/44) test

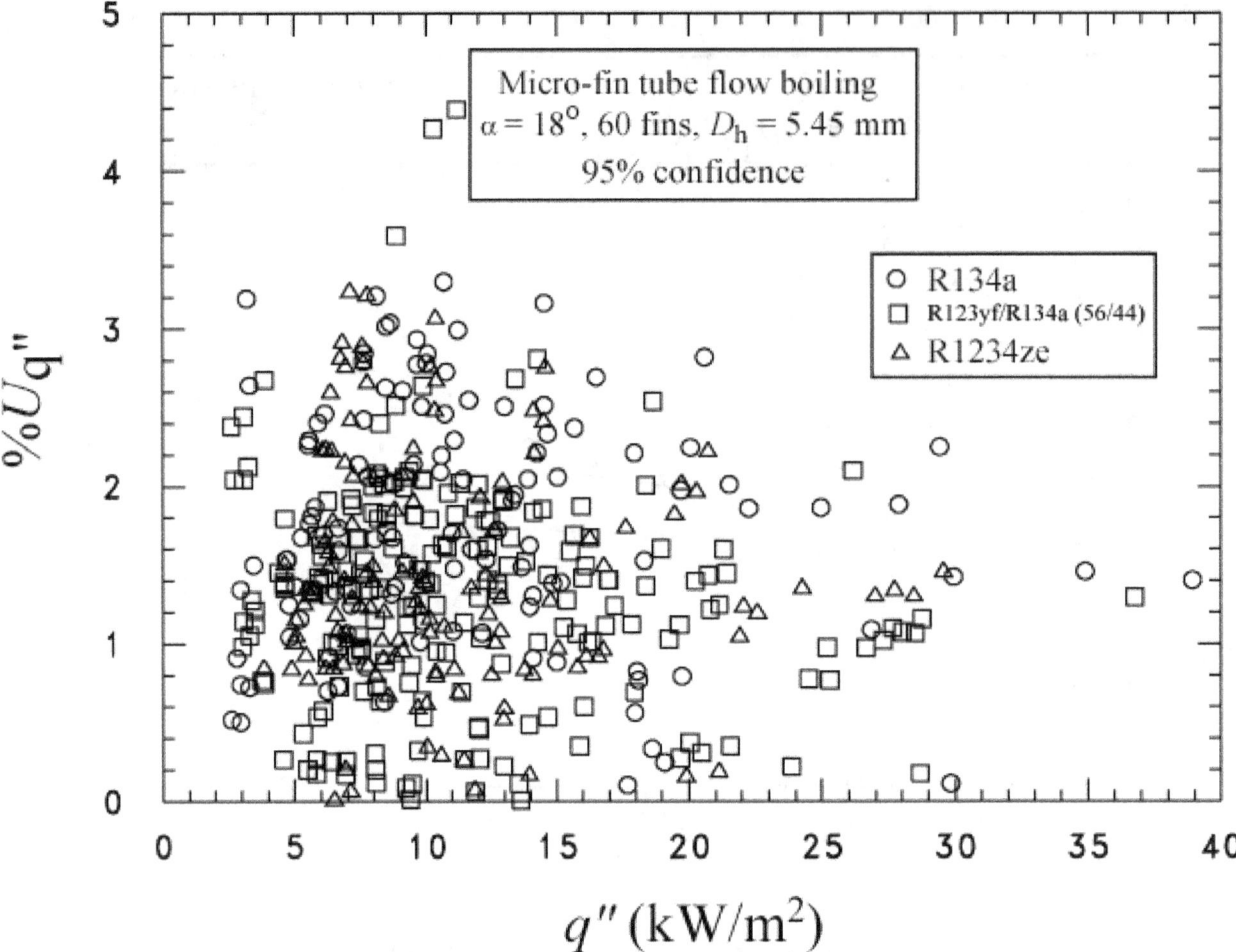

Figure 6 Relative uncertainty of water temperature gradient with respect to quality

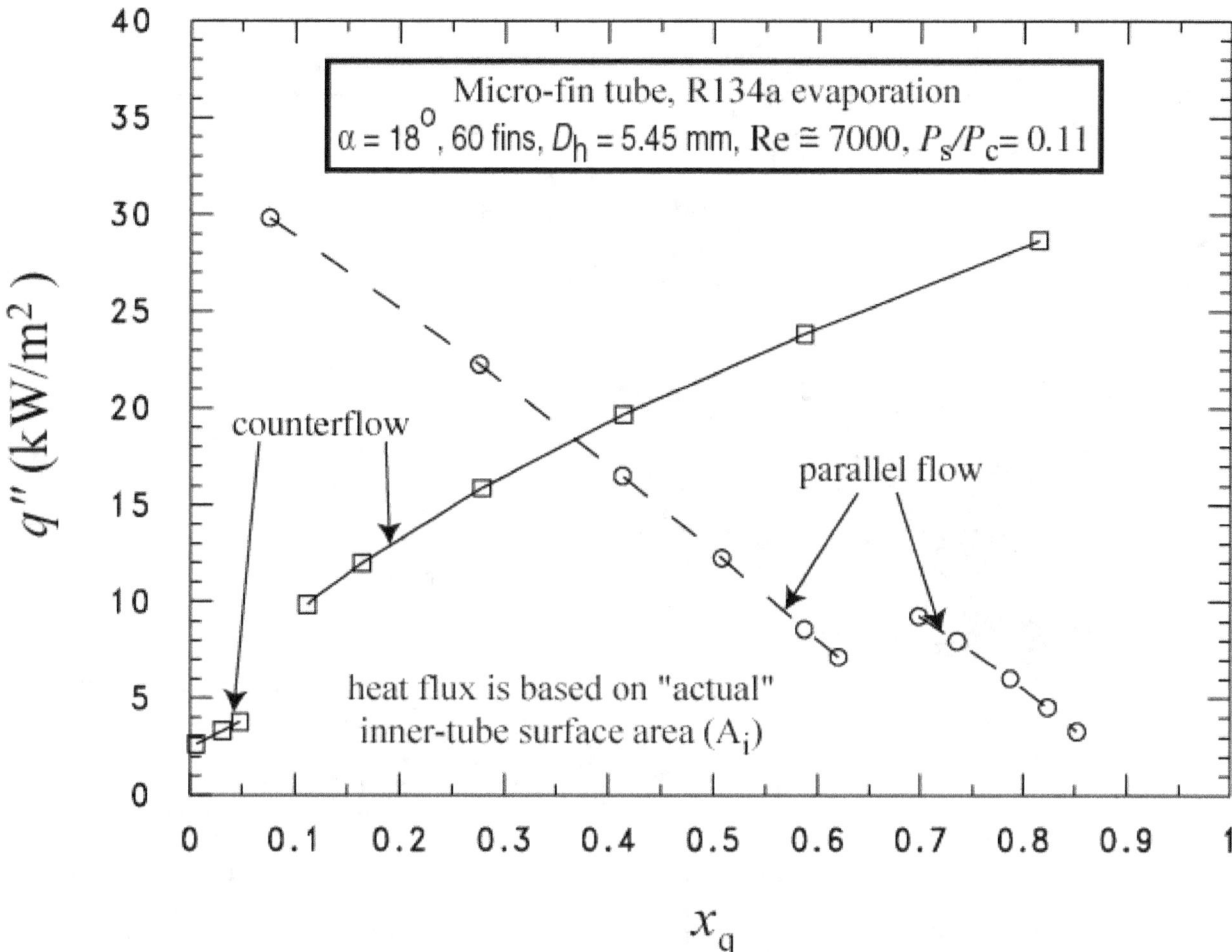

Figure 7 Heat Flux distribution for R134a

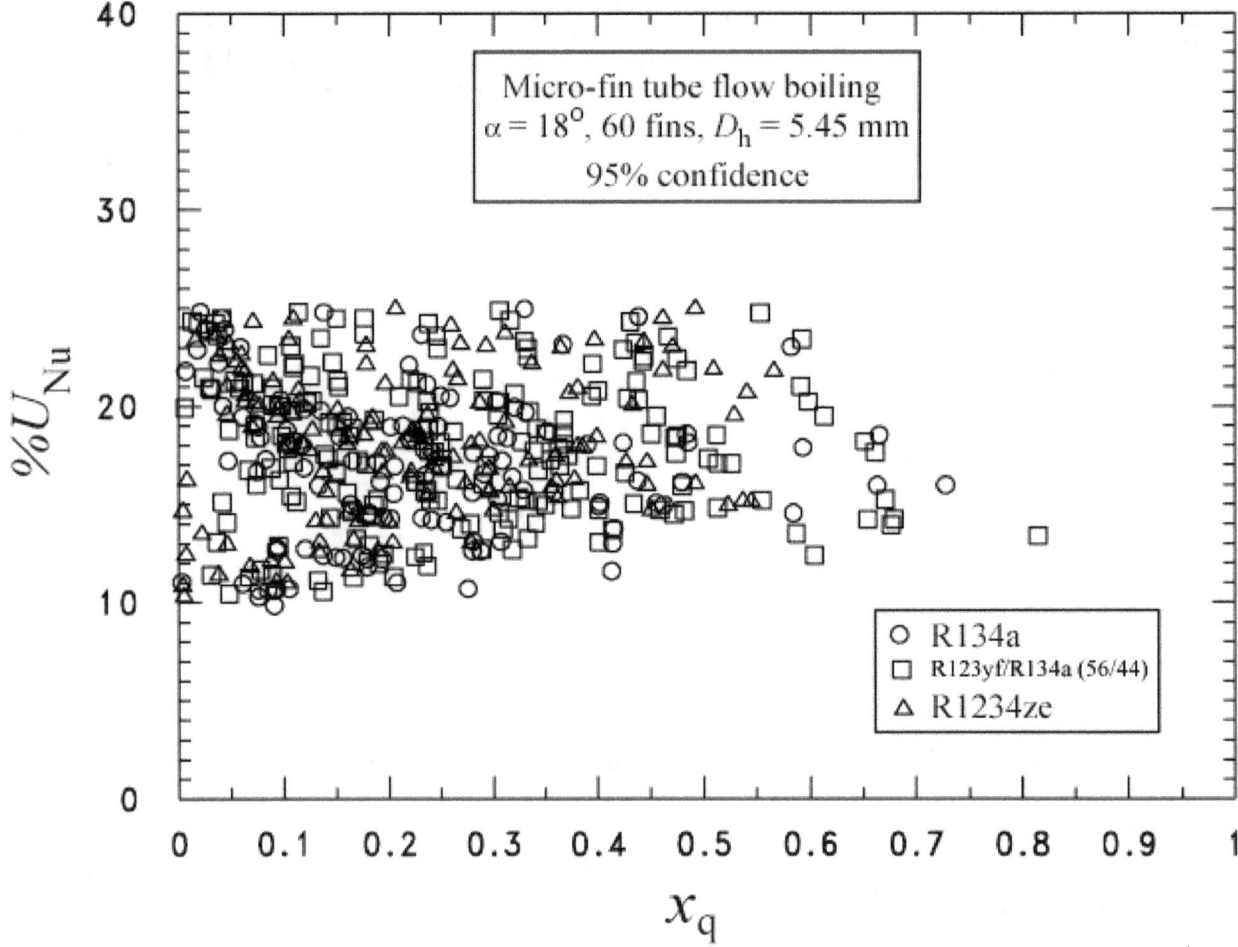

Figure 8 Relative uncertainty of the Nusselt number with respect to the quality

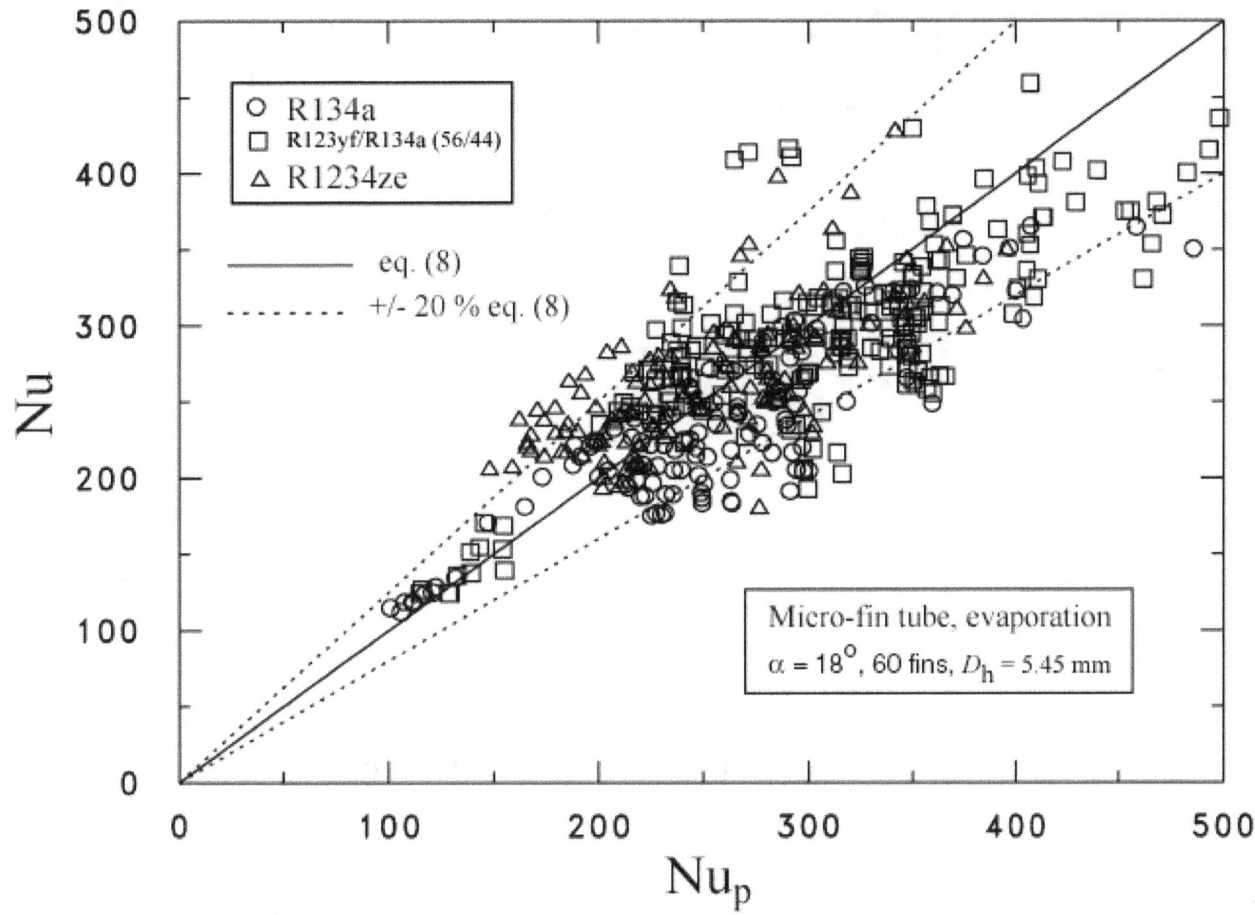

Figure 9 Comparison between measured Nusselt numbers and those predicted by the Hamilton et al. (2008) correlation

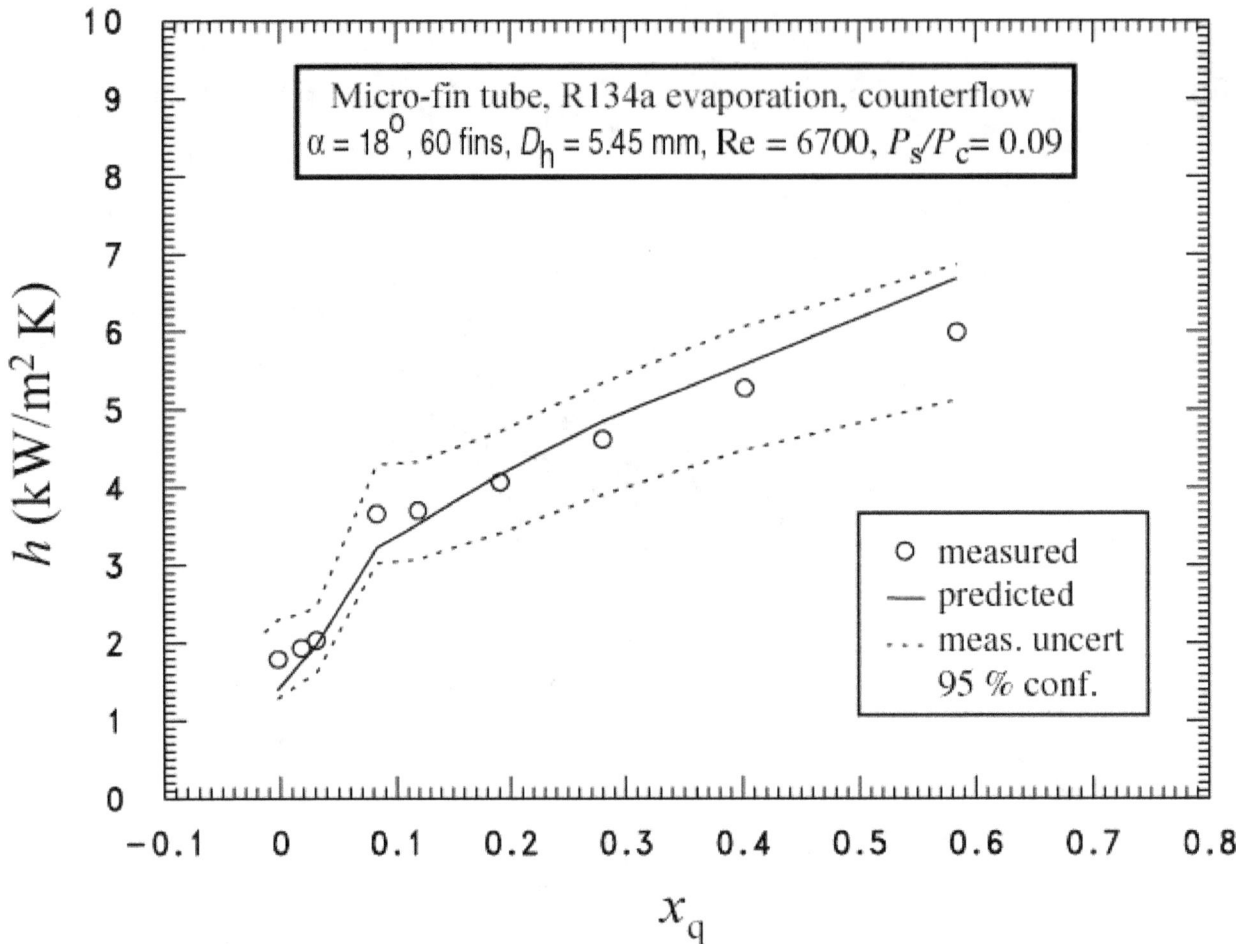

Figure 10 Flow boiling heat transfer coefficient for micro-fin tube versus thermodynamic quality for R134a

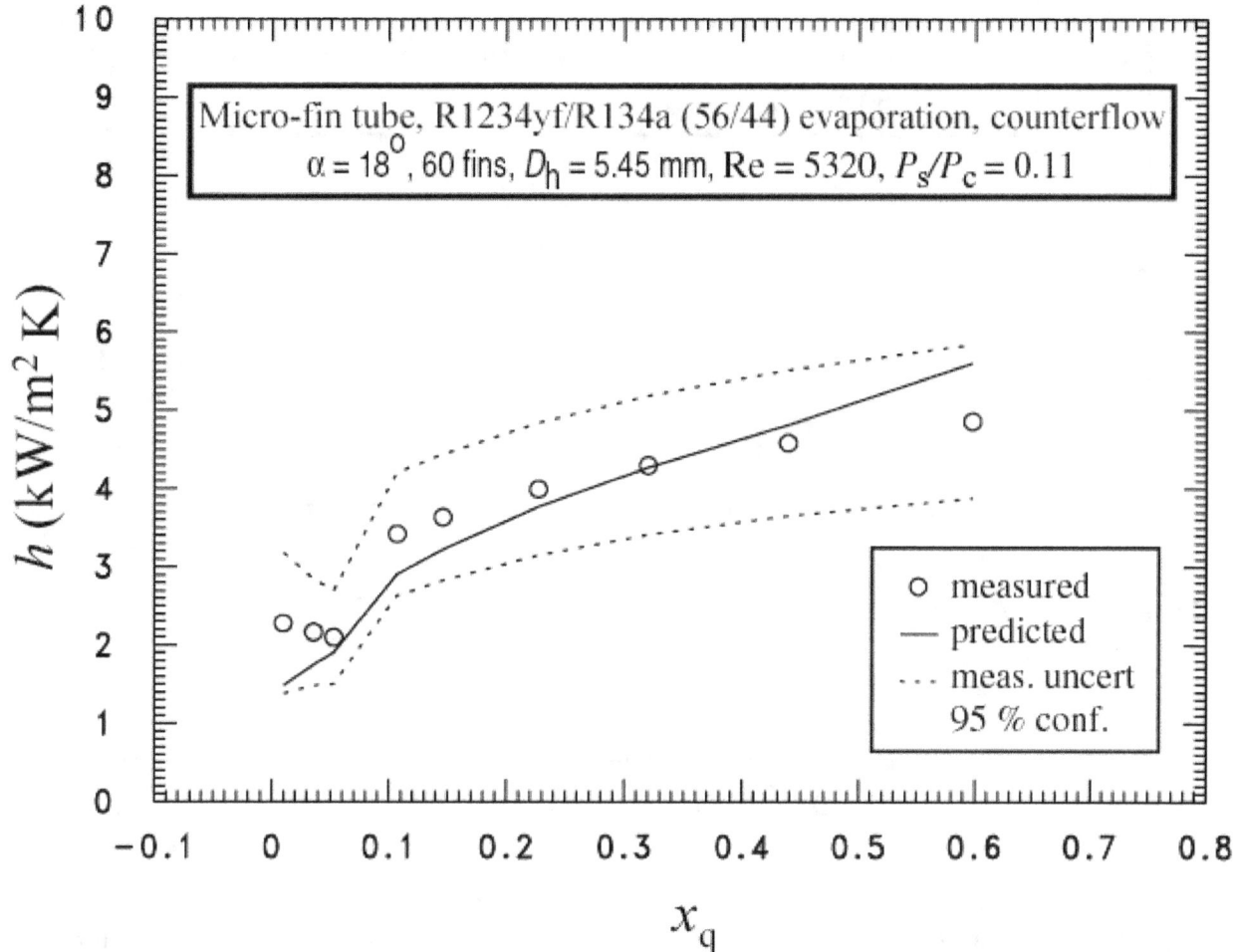

Figure 11 Flow boiling heat transfer coefficient for micro-fin tube versus thermodynamic quality for R1234yf/ R134a (56/44)

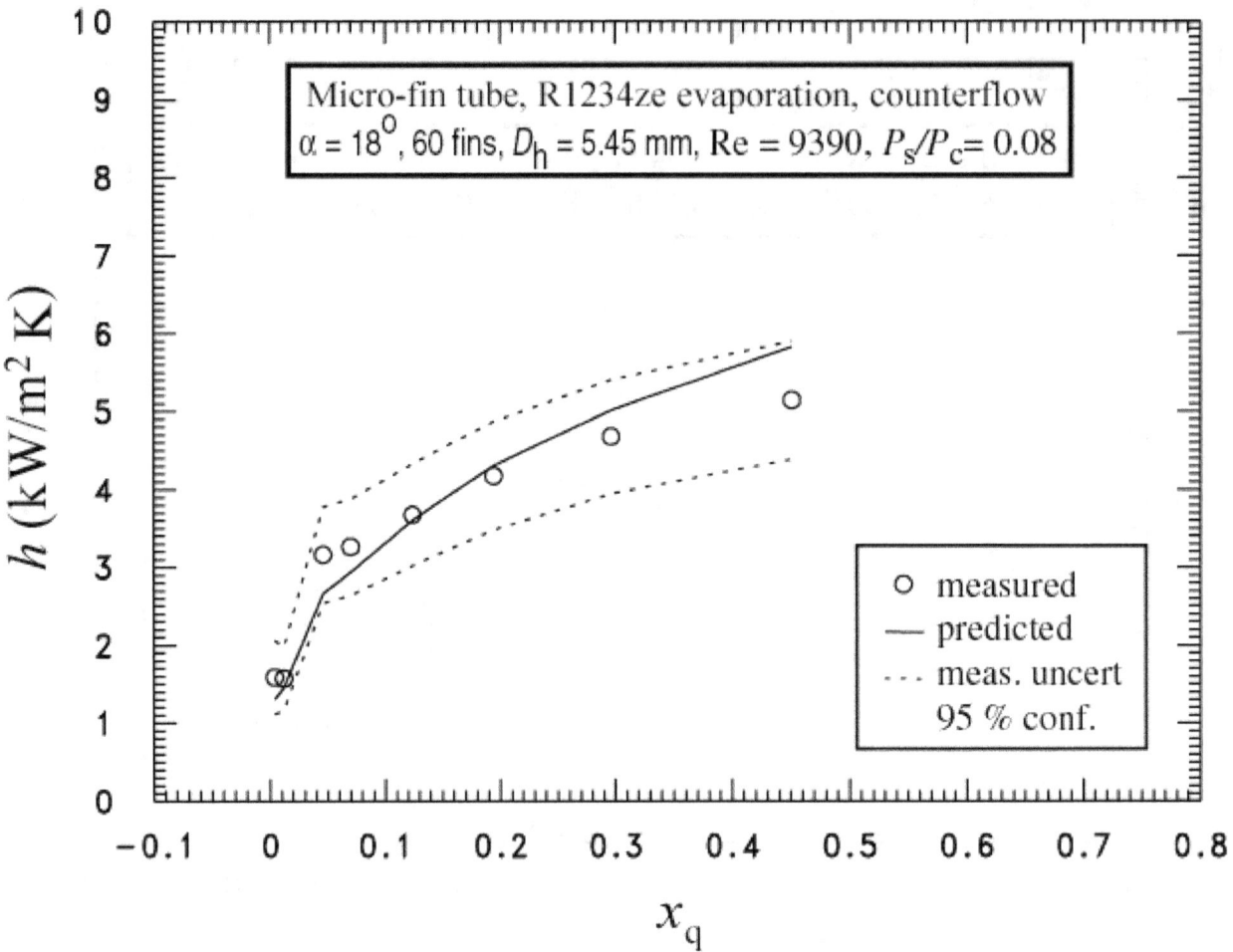

Figure 12 Flow boiling heat transfer coefficient for micro-fin tube versus thermodynamic quality for R1234ze(E) and counterflow

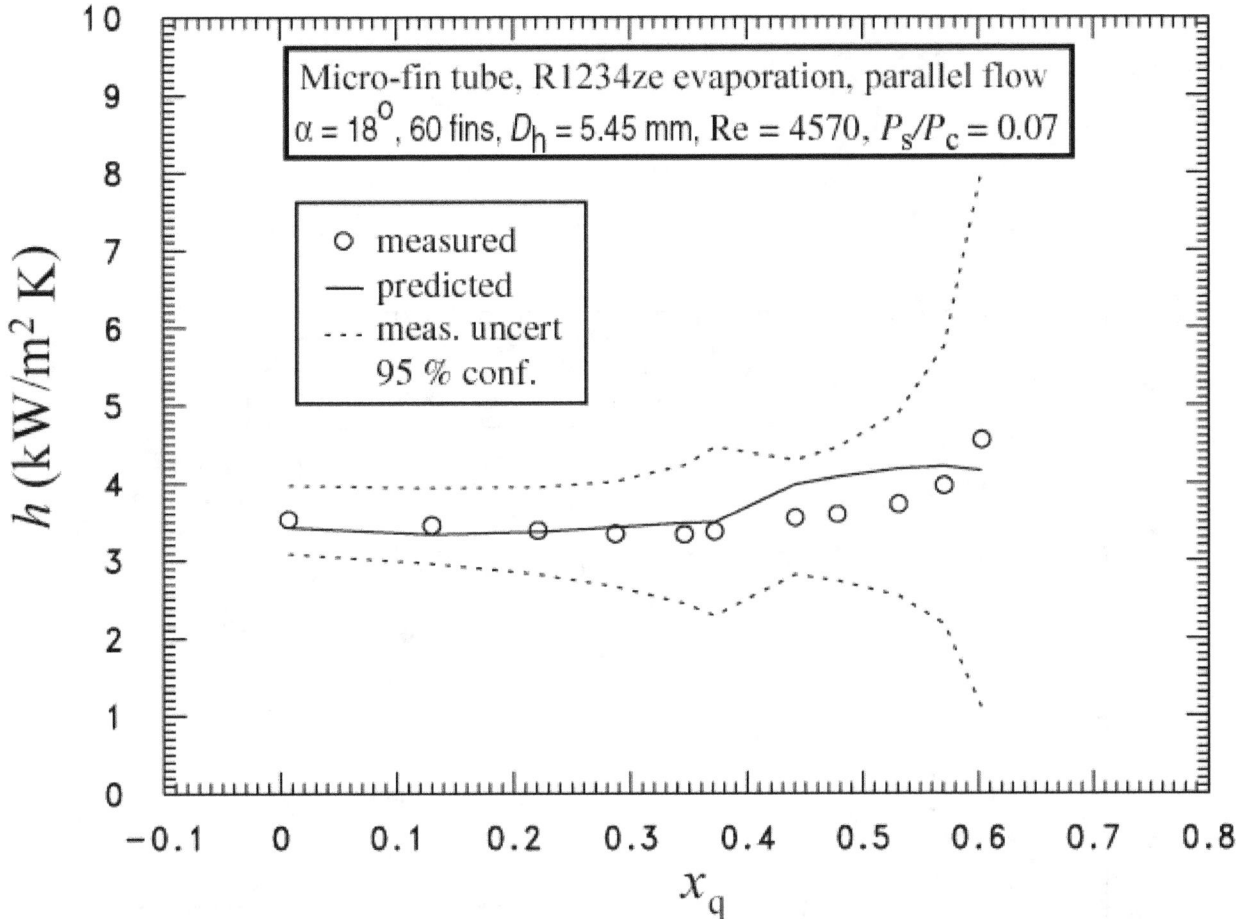

Figure 13 Flow boiling heat transfer coefficient for micro-fin tube versus thermodynamic quality for R1234ze(E) and parallel flow

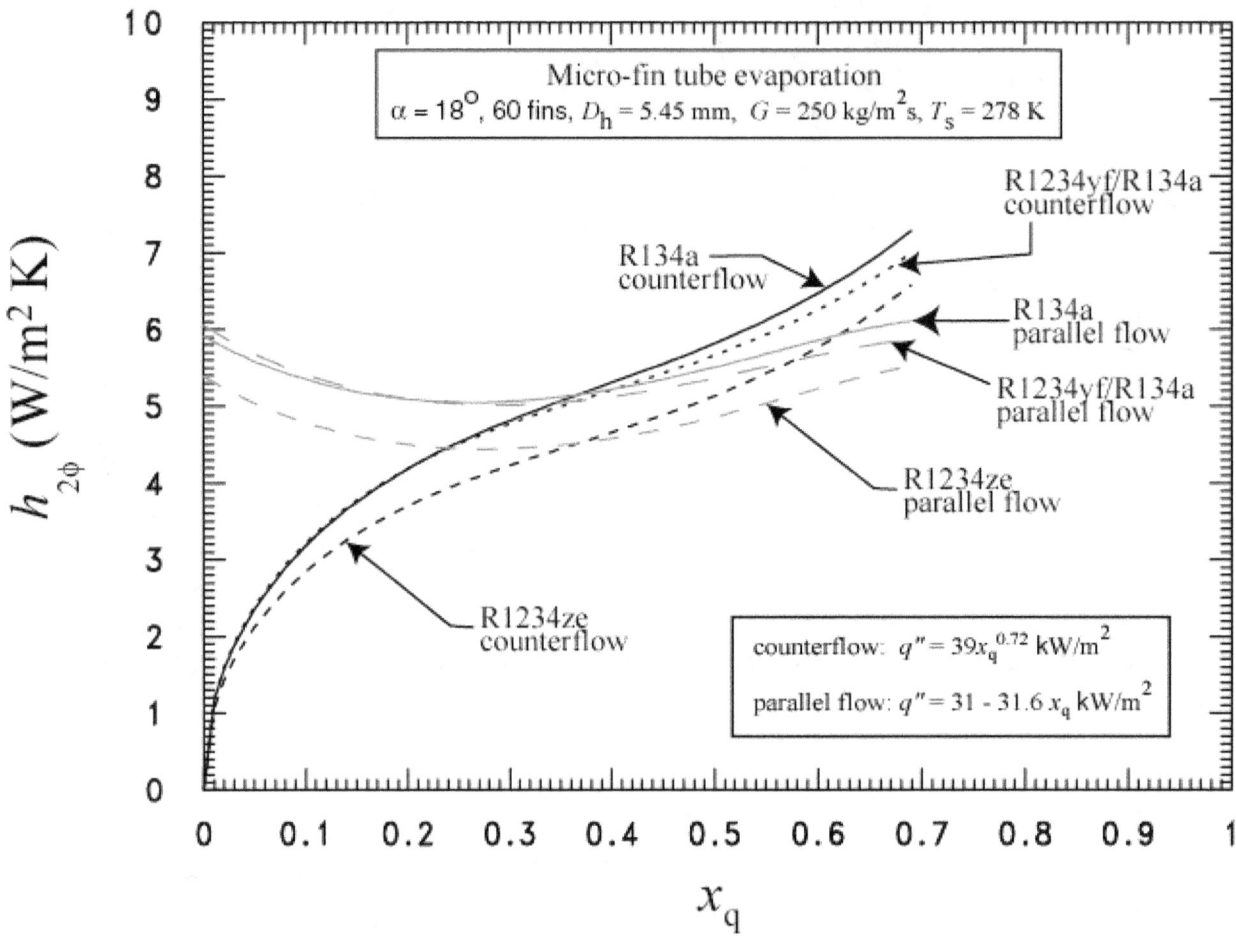

Figure 14 Flow boiling Nusselt numbers versus thermodynamic quality for test refrigerants

Convective Boiling of R134a within a micro-fin tube

(file: GWPNU.dat)

Nu	Re	x_q	Bo	P_s/P_c	T_s/T_c	M_w	S_v	Pr	flow	U_{Nu}
242.	7428.	0.11	0.24014×10^{-3}	0.097	0.753	102.030	8.280	3.70	P	10.7
218.	7411.	0.19	0.17939×10^{-3}	0.097	0.752	102.030	4.790	3.71	P	12.2
199.	7390.	0.26	0.13515×10^{-3}	0.096	0.752	102.030	3.700	3.71	P	14.1
185.	7360.	0.31	0.97960×10^{-4}	0.095	0.751	102.030	3.100	3.72	P	17.2
184.	7340.	0.33	0.83840×10^{-4}	0.094	0.750	102.030	2.890	3.72	P	19.7
235.	7522.	0.09	0.22350×10^{-3}	0.096	0.752	102.030	9.350	3.71	P	10.7
214.	7505.	0.17	0.16773×10^{-3}	0.095	0.751	102.030	5.290	3.71	P	12.4
196.	7486.	0.23	0.12687×10^{-3}	0.095	0.751	102.030	4.060	3.72	P	14.3
184.	7457.	0.28	0.92180×10^{-4}	0.094	0.750	102.030	3.380	3.72	P	17.6
183.	7439.	0.30	0.78780×10^{-4}	0.093	0.749	102.030	3.140	3.73	P	20.3
225.	8965.	0.09	0.17543×10^{-3}	0.095	0.751	102.030	9.420	3.71	P	10.6
205.	8946.	0.16	0.13211×10^{-3}	0.095	0.751	102.030	5.850	3.72	P	12.3
189.	8923.	0.20	0.10013×10^{-3}	0.094	0.750	102.030	4.620	3.72	P	14.3
177.	8889.	0.24	0.72620×10^{-4}	0.093	0.749	102.030	3.900	3.73	P	17.7
176.	8867.	0.26	0.61760×10^{-4}	0.093	0.749	102.030	3.650	3.73	P	20.4
225.	9023.	0.09	0.17280×10^{-3}	0.095	0.751	102.030	10.010	3.72	P	10.6
205.	9005.	0.15	0.13025×10^{-3}	0.095	0.751	102.030	6.110	3.72	P	12.3
189.	8982.	0.19	0.98760×10^{-4}	0.094	0.750	102.030	4.790	3.72	P	14.3
177.	8948.	0.23	0.71540×10^{-4}	0.093	0.749	102.030	4.040	3.73	P	17.8
175.	8926.	0.25	0.60730×10^{-4}	0.092	0.749	102.030	3.770	3.73	P	20.5
296.	4339.	0.33	0.21852×10^{-3}	0.088	0.745	102.030	2.910	3.76	C	25.0
303.	4329.	0.44	0.25595×10^{-3}	0.087	0.744	102.030	2.210	3.77	C	24.6
115.	6818.	0.03	0.43790×10^{-4}	0.090	0.746	102.030	24.850	3.75	C	23.9
119.	6814.	0.04	0.49460×10^{-4}	0.089	0.746	102.030	19.250	3.75	C	22.2
201.	6795.	0.08	0.10547×10^{-3}	0.089	0.746	102.030	10.870	3.76	C	18.4
214.	6793.	0.11	0.11917×10^{-3}	0.089	0.746	102.030	8.330	3.76	C	18.0
242.	6782.	0.17	0.15290×10^{-3}	0.088	0.745	102.030	5.600	3.76	C	17.2
270.	6765.	0.24	0.19596×10^{-3}	0.088	0.745	102.030	4.020	3.76	C	16.5
298.	6740.	0.33	0.25232×10^{-3}	0.087	0.744	102.030	2.920	3.77	C	15.7
324.	6702.	0.46	0.32798×10^{-3}	0.086	0.743	102.030	2.100	3.78	C	15.0
171.	9151.	0.04	0.71990×10^{-4}	0.089	0.746	102.030	17.880	3.75	C	20.0
181.	9147.	0.06	0.83020×10^{-4}	0.089	0.746	102.030	13.260	3.75	C	19.5
201.	9131.	0.10	0.10774×10^{-3}	0.089	0.746	102.030	8.560	3.76	C	18.8
218.	9106.	0.15	0.13739×10^{-3}	0.088	0.745	102.030	5.990	3.76	C	18.5
235.	9069.	0.22	0.17474×10^{-3}	0.087	0.744	102.030	4.290	3.77	C	18.4
250.	9015.	0.31	0.22355×10^{-3}	0.086	0.743	102.030	3.070	3.78	C	18.4
257.	4381.	0.09	0.46554×10^{-3}	0.089	0.746	102.030	9.300	3.76	P	12.7
223.	4372.	0.28	0.36102×10^{-3}	0.088	0.745	102.030	3.400	3.76	P	12.6
191.	4362.	0.41	0.27756×10^{-3}	0.088	0.745	102.030	2.340	3.76	P	13.0
230.	6817.	0.09	0.23034×10^{-3}	0.091	0.748	102.030	9.320	3.74	P	10.7

215.	6802.	0.18	0.17831×10^{-3}	0.091	0.747	102.030	5.180	3.74	P	12.2
202.	6784.	0.24	0.13868×10^{-3}	0.090	0.747	102.030	3.920	3.75	P	14.2
190.	6757.	0.30	0.10279×10^{-3}	0.089	0.746	102.030	3.230	3.75	P	17.5
187.	6739.	0.32	0.87500×10^{-4}	0.089	0.745	102.030	2.990	3.76	P	19.9
222.	9152.	0.08	0.16221×10^{-3}	0.094	0.750	102.030	10.980	3.72	P	10.7
208.	9133.	0.14	0.12518×10^{-3}	0.093	0.750	102.030	6.590	3.73	P	12.4
197.	9110.	0.18	0.97090×10^{-4}	0.093	0.749	102.030	5.120	3.73	P	14.6
188.	9077.	0.22	0.71830×10^{-4}	0.092	0.748	102.030	4.270	3.74	P	18.2
188.	9055.	0.24	0.61180×10^{-4}	0.091	0.748	102.030	3.970	3.74	P	21.1
129.	5054.	0.04	0.67940×10^{-4}	0.093	0.749	102.030	17.130	3.73	C	23.9
221.	5045.	0.10	0.14340×10^{-3}	0.093	0.749	102.030	8.920	3.73	C	20.4
233.	5044.	0.14	0.16033×10^{-3}	0.092	0.749	102.030	6.670	3.73	C	19.8
260.	5038.	0.21	0.20165×10^{-3}	0.092	0.749	102.030	4.400	3.73	C	19.0
290.	5030.	0.30	0.25414×10^{-3}	0.092	0.748	102.030	3.140	3.74	C	18.5
322.	5018.	0.42	0.32266×10^{-3}	0.091	0.748	102.030	2.280	3.74	C	18.1
357.	5001.	0.59	0.41459×10^{-3}	0.090	0.747	102.030	1.650	3.75	C	17.9
119.	6794.	0.02	0.50790×10^{-4}	0.093	0.750	102.030	30.860	3.73	C	22.9
125.	6791.	0.03	0.59270×10^{-4}	0.093	0.750	102.030	21.670	3.73	C	20.9
224.	6772.	0.08	0.13725×10^{-3}	0.093	0.749	102.030	10.220	3.73	C	17.3
227.	6769.	0.12	0.14923×10^{-3}	0.093	0.749	102.030	7.490	3.73	C	16.9
249.	6757.	0.19	0.18978×10^{-3}	0.092	0.748	102.030	4.850	3.73	C	16.2
283.	6738.	0.28	0.25027×10^{-3}	0.091	0.748	102.030	3.400	3.74	C	15.6
322.	6709.	0.40	0.33586×10^{-3}	0.090	0.747	102.030	2.400	3.75	C	15.1
365.	6667.	0.58	0.45673×10^{-3}	0.089	0.746	102.030	1.670	3.76	C	14.6
112.	9330.	0.01	0.42490×10^{-4}	0.099	0.754	102.030	44.540	3.69	C	21.8
209.	9301.	0.05	0.10841×10^{-3}	0.098	0.754	102.030	15.870	3.70	C	17.2
205.	9298.	0.08	0.11666×10^{-3}	0.098	0.754	102.030	10.980	3.70	C	16.7
220.	9279.	0.13	0.14929×10^{-3}	0.098	0.753	102.030	6.750	3.70	C	16.0
248.	9247.	0.20	0.20067×10^{-3}	0.097	0.752	102.030	4.570	3.70	C	15.5
284.	9198.	0.30	0.27508×10^{-3}	0.095	0.751	102.030	3.140	3.71	C	15.3
323.	9125.	0.46	0.38152×10^{-3}	0.093	0.750	102.030	2.120	3.73	C	15.1
259.	4620.	0.00	0.43607×10^{-3}	0.089	0.746	102.030	59.260	3.75	P	11.0
243.	4614.	0.18	0.34401×10^{-3}	0.089	0.746	102.030	5.190	3.75	P	11.8
229.	4606.	0.31	0.27101×10^{-3}	0.089	0.746	102.030	3.110	3.76	P	13.1
216.	4597.	0.40	0.21417×10^{-3}	0.088	0.745	102.030	2.410	3.76	P	14.9
205.	4583.	0.49	0.16087×10^{-3}	0.088	0.745	102.030	2.000	3.76	P	18.1
256.	7096.	0.09	0.31390×10^{-3}	0.102	0.756	102.030	9.300	3.67	P	9.8
234.	7080.	0.21	0.23759×10^{-3}	0.101	0.756	102.030	4.480	3.68	P	11.0
217.	7060.	0.29	0.18124×10^{-3}	0.100	0.755	102.030	3.290	3.68	P	12.6
204.	7031.	0.36	0.13275×10^{-3}	0.099	0.754	102.030	2.670	3.69	P	15.6
205.	7012.	0.39	0.11360×10^{-3}	0.099	0.754	102.030	2.460	3.69	P	18.1
222.	9148.	0.06	0.15780×10^{-3}	0.095	0.751	102.030	13.210	3.71	P	11.0
209.	9128.	0.12	0.12208×10^{-3}	0.095	0.751	102.030	7.420	3.72	P	12.7
199.	9105.	0.16	0.95190×10^{-4}	0.094	0.750	102.030	5.630	3.72	P	15.0
194.	9071.	0.20	0.71280×10^{-4}	0.093	0.749	102.030	4.630	3.73	P	19.0
196.	9049.	0.22	0.61370×10^{-4}	0.093	0.749	102.030	4.290	3.73	P	22.1
123.	4851.	0.04	0.66210×10^{-4}	0.090	0.747	102.030	17.960	3.75	C	24.5
134.	4849.	0.06	0.77800×10^{-4}	0.090	0.746	102.030	13.600	3.75	C	23.0
221.	4839.	0.12	0.15617×10^{-3}	0.089	0.746	102.030	7.540	3.75	C	19.9

239.	4837.	0.16	0.18009×10^{-3}	0.089	0.746	102.030	5.720	3.75	C	19.5
271.	4830.	0.25	0.22939×10^{-3}	0.089	0.746	102.030	3.810	3.76	C	19.0
299.	4822.	0.35	0.28478×10^{-3}	0.088	0.745	102.030	2.740	3.76	C	18.7
326.	4810.	0.48	0.35169×10^{-3}	0.088	0.745	102.030	2.010	3.76	C	18.6
351.	4793.	0.66	0.43643×10^{-3}	0.087	0.744	102.030	1.470	3.77	C	18.5
118.	7039.	0.02	0.52890×10^{-4}	0.093	0.749	102.030	28.360	3.73	C	24.8
224.	7024.	0.07	0.14211×10^{-3}	0.092	0.749	102.030	11.600	3.73	C	19.2
224.	7022.	0.11	0.15874×10^{-3}	0.092	0.749	102.030	8.090	3.73	C	18.3
246.	7009.	0.19	0.21377×10^{-3}	0.092	0.748	102.030	4.930	3.74	C	17.2
281.	6986.	0.29	0.29511×10^{-3}	0.091	0.748	102.030	3.280	3.74	C	16.6
322.	6950.	0.44	0.40972×10^{-3}	0.090	0.746	102.030	2.210	3.75	C	16.2
365.	6897.	0.66	0.57110×10^{-3}	0.088	0.745	102.030	1.480	3.76	C	16.0
243.	7199.	0.08	0.15722×10^{-3}	0.095	0.751	102.030	10.990	3.72	C	19.0
241.	7196.	0.12	0.17570×10^{-3}	0.095	0.751	102.030	7.550	3.72	C	18.0
265.	7181.	0.21	0.23646×10^{-3}	0.094	0.750	102.030	4.540	3.72	C	16.9
302.	7156.	0.32	0.32609×10^{-3}	0.093	0.750	102.030	3.000	3.73	C	16.4
346.	7119.	0.48	0.45225×10^{-3}	0.092	0.749	102.030	2.020	3.73	C	16.1
393.	7065.	0.73	0.62981×10^{-3}	0.090	0.747	102.030	1.350	3.75	C	16.0
238.	9743.	0.14	0.19109×10^{-3}	0.113	0.765	102.030	6.400	3.62	C	24.8
266.	9713.	0.23	0.27039×10^{-3}	0.112	0.764	102.030	4.020	3.62	C	23.6
304.	9668.	0.37	0.38578×10^{-3}	0.111	0.763	102.030	2.600	3.63	C	23.1
350.	9599.	0.58	0.55143×10^{-3}	0.109	0.762	102.030	1.670	3.64	C	23.0
279.	4638.	0.09	0.46715×10^{-3}	0.089	0.746	102.030	9.320	3.76	P	12.8
250.	4629.	0.28	0.36310×10^{-3}	0.088	0.745	102.030	3.390	3.76	P	13.1
220.	4618.	0.42	0.27948×10^{-3}	0.088	0.745	102.030	2.330	3.76	P	13.8
320.	7314.	0.08	0.50762×10^{-3}	0.107	0.760	102.030	10.770	3.65	P	10.3
282.	7297.	0.28	0.37854×10^{-3}	0.106	0.760	102.030	3.420	3.65	P	10.7
249.	7278.	0.41	0.28030×10^{-3}	0.105	0.759	102.030	2.320	3.65	P	11.6

APPENDIX A2

Convective Boiling of R1234yf/ R134a (56/44) within a micro-fin tube

(file: GWPNU.dat)

Nu	Re	x_q	Bo	P_s/P_c	T_s/T_c	M_w	Sv	Pr	flow	U_{Nu}
284.	5241.	0.18	0.16094×10^{-3}	0.107	0.758	108.910	5.200	3.85	C	24.5
302.	5236.	0.25	0.19118×10^{-3}	0.107	0.757	108.910	3.800	3.86	C	23.6
312.	5228.	0.33	0.22311×10^{-3}	0.106	0.757	108.910	2.860	3.86	C	22.9
316.	5217.	0.44	0.25893×10^{-3}	0.106	0.757	108.910	2.170	3.86	C	22.5
154.	5801.	0.03	0.77280×10^{-4}	0.111	0.761	108.910	18.980	3.84	C	24.2
264.	5790.	0.10	0.16835×10^{-3}	0.110	0.760	108.910	8.690	3.84	C	20.0
273.	5788.	0.14	0.18871×10^{-3}	0.110	0.760	108.910	6.260	3.84	C	19.2
295.	5781.	0.23	0.23489×10^{-3}	0.110	0.760	108.910	3.990	3.84	C	18.2
319.	5771.	0.34	0.29078×10^{-3}	0.109	0.759	108.910	2.810	3.84	C	17.8
346.	5758.	0.47	0.36167×10^{-3}	0.109	0.759	108.910	2.040	3.85	C	17.6
375.	5740.	0.66	0.45483×10^{-3}	0.108	0.758	108.910	1.480	3.85	C	17.6
269.	7971.	0.09	0.11760×10^{-3}	0.115	0.764	108.910	9.230	3.82	C	20.9
280.	7966.	0.12	0.13196×10^{-3}	0.115	0.764	108.910	7.160	3.82	C	20.2
307.	7955.	0.19	0.16690×10^{-3}	0.115	0.763	108.910	4.910	3.82	C	19.3
339.	7938.	0.26	0.21118×10^{-3}	0.114	0.763	108.910	3.570	3.82	C	18.7
372.	7915.	0.36	0.26892×10^{-3}	0.113	0.762	108.910	2.630	3.83	C	17.9
408.	7883.	0.50	0.34626×10^{-3}	0.112	0.761	108.910	1.920	3.83	C	17.3
137.	7708.	0.03	0.62980×10^{-4}	0.110	0.760	108.910	20.720	3.84	C	20.8
139.	7702.	0.05	0.71900×10^{-4}	0.109	0.759	108.910	15.340	3.84	C	18.8
271.	7672.	0.11	0.18684×10^{-3}	0.108	0.759	108.910	7.750	3.85	C	15.1
302.	7665.	0.16	0.22756×10^{-3}	0.108	0.758	108.910	5.520	3.85	C	14.6
341.	7646.	0.28	0.30068×10^{-3}	0.107	0.758	108.910	3.390	3.85	C	14.0
363.	7619.	0.41	0.37259×10^{-3}	0.106	0.757	108.910	2.320	3.86	C	13.7
375.	7583.	0.59	0.45080×10^{-3}	0.105	0.756	108.910	1.660	3.87	C	13.5
380.	7531.	0.82	0.54107×10^{-3}	0.103	0.754	108.910	1.200	3.88	C	13.4
127.	10779.	0.01	0.38460×10^{-4}	0.117	0.765	108.910	31.760	3.81	C	24.3
124.	10773.	0.02	0.43420×10^{-4}	0.117	0.765	108.910	23.890	3.81	C	21.5
238.	10731.	0.07	0.11483×10^{-3}	0.116	0.764	108.910	11.940	3.81	C	16.8
244.	10725.	0.10	0.12879×10^{-3}	0.116	0.764	108.910	8.750	3.81	C	16.2
269.	10701.	0.16	0.16851×10^{-3}	0.115	0.764	108.910	5.610	3.82	C	15.6
301.	10661.	0.24	0.22343×10^{-3}	0.114	0.763	108.910	3.880	3.82	C	15.2
336.	10603.	0.35	0.29836×10^{-3}	0.112	0.762	108.910	2.720	3.83	C	15.0
372.	10518.	0.51	0.40169×10^{-3}	0.110	0.760	108.910	1.890	3.84	C	14.8
297.	5294.	0.04	0.27103×10^{-3}	0.118	0.765	108.910	16.970	3.81	P	15.1
274.	5291.	0.15	0.21608×10^{-3}	0.118	0.765	108.910	5.930	3.81	P	16.7
251.	5286.	0.23	0.17055×10^{-3}	0.117	0.765	108.910	4.010	3.81	P	18.8
227.	5281.	0.29	0.13316×10^{-3}	0.117	0.765	108.910	3.250	3.81	P	21.4
309.	5720.	0.19	0.38506×10^{-3}	0.115	0.764	108.910	4.810	3.82	P	12.0
283.	5711.	0.33	0.29722×10^{-3}	0.115	0.763	108.910	2.850	3.82	P	13.3
261.	5700.	0.43	0.22985×10^{-3}	0.114	0.763	108.910	2.210	3.82	P	15.0
323.	7789.	0.17	0.33194×10^{-3}	0.118	0.766	108.910	5.420	3.80	P	11.3

300.	7771.	0.29	0.25308×10^{-3}	0.117	0.765	108.910	3.260	3.81	P	12.7
281.	7750.	0.38	0.19347×10^{-3}	0.117	0.765	108.910	2.540	3.81	P	14.8
267.	7719.	0.45	0.14009×10^{-3}	0.115	0.764	108.910	2.130	3.82	P	18.6
266.	7699.	0.48	0.11772×10^{-3}	0.115	0.763	108.910	1.990	3.82	P	21.8
310.	10387.	0.09	0.24932×10^{-3}	0.119	0.766	108.910	9.300	3.80	P	11.4
292.	10359.	0.18	0.18967×10^{-3}	0.118	0.766	108.910	5.000	3.80	P	12.9
279.	10326.	0.25	0.14474×10^{-3}	0.117	0.765	108.910	3.770	3.81	P	15.2
273.	10278.	0.30	0.10475×10^{-3}	0.116	0.764	108.910	3.100	3.81	P	19.5
279.	10246.	0.33	0.88140×10^{-4}	0.115	0.763	108.910	2.880	3.82	P	23.3
244.	5469.	0.09	0.15067×10^{-3}	0.110	0.760	108.910	9.830	3.84	C	22.6
255.	5467.	0.12	0.17290×10^{-3}	0.110	0.760	108.910	7.030	3.84	C	21.6
274.	5462.	0.21	0.21700×10^{-3}	0.109	0.759	108.910	4.430	3.84	C	20.5
290.	5454.	0.31	0.26493×10^{-3}	0.109	0.759	108.910	3.100	3.85	C	20.2
305.	5443.	0.43	0.32148×10^{-3}	0.108	0.759	108.910	2.250	3.85	C	20.4
319.	5428.	0.59	0.39173×10^{-3}	0.108	0.758	108.910	1.650	3.85	C	21.0
151.	7778.	0.04	0.58980×10^{-4}	0.113	0.762	108.910	18.410	3.83	C	23.6
153.	7774.	0.05	0.67080×10^{-4}	0.113	0.762	108.910	14.270	3.83	C	21.3
265.	7751.	0.11	0.15365×10^{-3}	0.112	0.761	108.910	7.920	3.83	C	17.1
282.	7747.	0.15	0.17672×10^{-3}	0.112	0.761	108.910	5.970	3.83	C	16.6
313.	7733.	0.24	0.22931×10^{-3}	0.111	0.761	108.910	3.930	3.84	C	15.8
343.	7711.	0.34	0.29310×10^{-3}	0.110	0.760	108.910	2.780	3.84	C	15.2
371.	7678.	0.48	0.37406×10^{-3}	0.109	0.759	108.910	P0	3.85	C	14.7
397.	7631.	0.68	0.48040×10^{-3}	0.107	0.758	108.910	1.440	3.85	C	14.2
244.	10387.	0.06	0.11735×10^{-3}	0.112	0.761	108.910	13.320	3.83	C	20.9
247.	10382.	0.09	0.13363×10^{-3}	0.112	0.761	108.910	9.390	3.83	C	20.1
264.	10359.	0.16	0.17514×10^{-3}	0.111	0.761	108.910	5.780	3.83	C	19.2
285.	10320.	0.24	0.22921×10^{-3}	0.110	0.760	108.910	3.920	3.84	C	18.8
308.	10262.	0.35	0.30072×10^{-3}	0.109	0.759	108.910	2.730	3.85	C	18.6
330.	10177.	0.51	0.39729×10^{-3}	0.106	0.757	108.910	1.890	3.86	C	18.5
124.	10683.	0.02	0.36850×10^{-4}	0.115	0.764	108.910	26.920	3.82	C	24.2
124.	10675.	0.03	0.43370×10^{-4}	0.115	0.764	108.910	21.110	3.82	C	21.0
250.	10633.	0.07	0.12420×10^{-3}	0.114	0.763	108.910	10.840	3.82	C	16.0
254.	10627.	0.11	0.13885×10^{-3}	0.114	0.763	108.910	8.000	3.82	C	15.4
286.	10601.	0.18	0.18831×10^{-3}	0.113	0.762	108.910	5.120	3.83	C	14.5
331.	10556.	0.27	0.26199×10^{-3}	0.112	0.761	108.910	3.490	3.83	C	13.8
381.	10489.	0.40	0.36610×10^{-3}	0.110	0.760	108.910	2.390	3.84	C	13.1
432.	10390.	0.60	0.51285×10^{-3}	0.107	0.758	108.910	1.610	3.85	C	12.4
315.	5274.	0.04	0.36571×10^{-3}	0.113	0.762	108.910	18.710	3.83	P	13.1
290.	5269.	0.18	0.28783×10^{-3}	0.113	0.762	108.910	4.980	3.83	P	14.3
266.	5263.	0.29	0.22567×10^{-3}	0.113	0.762	108.910	3.250	3.83	P	16.0
243.	5255.	0.37	0.17686×10^{-3}	0.112	0.761	108.910	2.590	3.83	P	18.1
216.	5243.	0.44	0.13051×10^{-3}	0.112	0.761	108.910	2.200	3.83	P	21.3
202.	5235.	0.47	0.10952×10^{-3}	0.111	0.761	108.910	2.060	3.84	P	23.5
310.	10387.	0.09	0.24932×10^{-3}	0.119	0.766	108.910	9.300	3.80	P	11.4
292.	10359.	0.18	0.18967×10^{-3}	0.118	0.766	108.910	5.000	3.80	P	12.9
279.	10326.	0.25	0.14474×10^{-3}	0.117	0.765	108.910	3.770	3.81	P	15.2
273.	10278.	0.30	0.10475×10^{-3}	0.116	0.764	108.910	3.100	3.81	P	19.5
279.	10246.	0.33	0.88140×10^{-4}	0.115	0.763	108.910	2.880	3.82	P	23.3
288.	10501.	0.14	0.26964×10^{-3}	0.121	0.767	108.910	6.430	3.79	P	10.6

272.	10470.	0.24	0.20654×10^{-3}	0.120	0.767	108.910	3.910	3.80	P	11.9
261.	10433.	0.31	0.15896×10^{-3}	0.119	0.766	108.910	3.050	3.80	P	13.7
258.	10380.	0.37	0.11651×10^{-3}	0.117	0.765	108.910	2.560	3.81	P	17.5
266.	10346.	0.40	0.98810×10^{-4}	0.116	0.764	108.910	2.390	3.81	P	20.8
271.	5472.	0.11	0.15917×10^{-3}	0.112	0.762	108.910	7.890	3.83	C	22.0
287.	5470.	0.15	0.18119×10^{-3}	0.112	0.762	108.910	5.910	3.83	C	21.0
316.	5465.	0.24	0.22800×10^{-3}	0.112	0.761	108.910	3.890	3.83	C	19.7
345.	5457.	0.34	0.28194×10^{-3}	0.112	0.761	108.910	2.790	3.83	C	18.9
373.	5446.	0.47	0.34824×10^{-3}	0.111	0.761	108.910	2.040	3.84	C	18.4
402.	5431.	0.65	0.43337×10^{-3}	0.110	0.760	108.910	1.500	3.84	C	18.2
169.	7830.	0.04	0.68220×10^{-4}	0.117	0.765	108.910	15.720	3.81	C	22.9
284.	7810.	0.10	0.15178×10^{-3}	0.116	0.764	108.910	8.300	3.81	C	18.1
291.	7807.	0.14	0.16866×10^{-3}	0.116	0.764	108.910	6.230	3.81	C	17.3
318.	7794.	0.23	0.21545×10^{-3}	0.116	0.764	108.910	4.120	3.81	C	16.2
353.	7775.	0.32	0.27933×10^{-3}	0.115	0.764	108.910	2.930	3.82	C	15.3
393.	7746.	0.46	0.36599×10^{-3}	0.114	0.763	108.910	2.100	3.82	C	14.7
436.	7705.	0.65	0.48519×10^{-3}	0.112	0.762	108.910	1.490	3.83	C	14.2
136.	10224.	0.03	0.47430×10^{-4}	0.114	0.763	108.910	21.440	3.82	C	23.8
241.	10188.	0.07	0.11838×10^{-3}	0.113	0.762	108.910	11.100	3.83	C	18.4
247.	10183.	0.10	0.13340×10^{-3}	0.113	0.762	108.910	8.220	3.83	C	17.8
268.	10161.	0.17	0.17380×10^{-3}	0.112	0.761	108.910	5.320	3.83	C	17.2
295.	10125.	0.25	0.22807×10^{-3}	0.111	0.760	108.910	3.710	3.84	C	17.0
324.	10071.	0.36	0.30102×10^{-3}	0.109	0.759	108.910	2.630	3.84	C	17.0
354.	9993.	0.53	0.40066×10^{-3}	0.107	0.758	108.910	1.840	3.86	C	17.1
270.	5422.	0.09	0.20465×10^{-3}	0.112	0.761	108.910	9.500	3.83	P	16.8
247.	5417.	0.17	0.16173×10^{-3}	0.112	0.761	108.910	5.500	3.83	P	18.9
223.	5412.	0.22	0.12630×10^{-3}	0.111	0.761	108.910	4.210	3.84	P	21.4
274.	5322.	0.14	0.28379×10^{-3}	0.110	0.760	108.910	6.200	3.84	P	14.3
253.	5315.	0.25	0.22243×10^{-3}	0.109	0.759	108.910	3.750	3.84	P	16.9
231.	5307.	0.33	0.17402×10^{-3}	0.109	0.759	108.910	2.910	3.85	P	18.2
206.	5295.	0.39	0.12773×10^{-3}	0.108	0.759	108.910	2.430	3.85	P	20.5
193.	5286.	0.42	0.10660×10^{-3}	0.108	0.758	108.910	2.270	3.85	P	22.9
298.	7573.	0.05	0.38678×10^{-3}	0.109	0.759	108.910	15.420	3.85	P	10.5
285.	7558.	0.20	0.30408×10^{-3}	0.108	0.759	108.910	4.520	3.85	P	11.3
273.	7539.	0.32	0.23822×10^{-3}	0.108	0.758	108.910	2.990	3.85	P	12.7
263.	7515.	0.40	0.18667×10^{-3}	0.107	0.757	108.910	2.390	3.86	P	14.7
255.	7480.	0.48	0.13796×10^{-3}	0.105	0.756	108.910	2.040	3.86	P	18.5
313.	10773.	0.08	0.24388×10^{-3}	0.119	0.766	108.910	10.360	3.80	P	11.7
300.	10744.	0.17	0.18725×10^{-3}	0.118	0.766	108.910	5.330	3.80	P	13.3
290.	10708.	0.23	0.14371×10^{-3}	0.117	0.765	108.910	3.960	3.81	P	15.6
286.	10656.	0.29	0.10366×10^{-3}	0.116	0.764	108.910	3.230	3.82	P	20.3
291.	10622.	0.32	0.86250×10^{-4}	0.115	0.763	108.910	2.990	3.82	P	24.4
396.	5319.	0.55	0.32278×10^{-3}	0.107	0.757	108.910	1.760	3.86	C	24.7
313.	7630.	0.11	0.15615×10^{-3}	0.107	0.758	108.910	8.010	3.85	C	19.8
329.	7626.	0.15	0.17889×10^{-3}	0.107	0.758	108.910	6.000	3.86	C	19.0
355.	7611.	0.24	0.22937×10^{-3}	0.106	0.757	108.910	3.930	3.86	C	17.7
378.	7587.	0.34	0.28945×10^{-3}	0.106	0.756	108.910	2.780	3.86	C	16.7
398.	7553.	0.48	0.36480×10^{-3}	0.104	0.756	108.910	2.020	3.87	C	15.9
415.	7504.	0.67	0.46293×10^{-3}	0.102	0.754	108.910	1.460	3.88	C	15.2

289.	10135.	0.07	0.12690×10^{-3}	0.117	0.765	108.910	10.930	3.81	C	19.0
294.	10129.	0.11	0.14293×10^{-3}	0.117	0.765	108.910	7.980	3.81	C	18.2
315.	107.	0.18	0.18559×10^{-3}	0.116	0.764	108.910	5.100	3.81	C	17.1
342.	10072.	0.27	0.24253×10^{-3}	0.115	0.763	108.910	3.540	3.82	C	16.3
371.	10019.	0.38	0.31883×10^{-3}	0.113	0.762	108.910	2.500	3.83	C	15.7
400.	9943.	0.56	0.42286×10^{-3}	0.111	0.761	108.910	1.750	3.84	C	15.2
339.	5309.	0.07	0.20810×10^{-3}	0.116	0.764	108.910	11.060	3.81	P	21.2
315.	5305.	0.15	0.16269×10^{-3}	0.116	0.764	108.910	5.960	3.82	P	24.5
331.	7761.	0.03	0.40682×10^{-3}	0.110	0.760	108.910	21.010	3.84	P	11.4
321.	7745.	0.20	0.31810×10^{-3}	0.109	0.759	108.910	4.730	3.85	P	12.5
313.	7723.	0.31	0.24761×10^{-3}	0.108	0.759	108.910	3.030	3.85	P	14.2
309.	7697.	0.40	0.19261×10^{-3}	0.107	0.758	108.910	2.400	3.85	P	16.9
313.	7659.	0.48	0.14084×10^{-3}	0.106	0.757	108.910	2.030	3.86	P	22.4
341.	10648.	0.10	0.23945×10^{-3}	0.124	0.770	108.910	8.720	3.78	P	12.9
336.	10622.	0.18	0.18602×10^{-3}	0.123	0.769	108.910	4.890	3.78	P	15.0
335.	10590.	0.25	0.14354×10^{-3}	0.122	0.768	108.910	3.720	3.79	P	18.1
344.	10544.	0.31	0.10243×10^{-3}	0.121	0.768	108.910	3.070	3.79	P	24.9
430.	5182.	0.43	0.31165×10^{-3}	0.114	0.763	108.910	2.230	3.82	C	24.3
459.	5171.	0.59	0.39198×10^{-3}	0.113	0.762	108.910	1.640	3.83	C	23.4
170.	7750.	0.04	0.64350×10^{-4}	0.111	0.760	108.910	17.280	3.84	C	24.5
299.	7724.	0.10	0.15596×10^{-3}	0.110	0.760	108.910	8.730	3.84	C	18.5
308.	7719.	0.14	0.17608×10^{-3}	0.110	0.760	108.910	6.400	3.84	C	17.6
336.	7704.	0.23	0.22740×10^{-3}	0.109	0.759	108.910	4.120	3.85	C	16.2
368.	7681.	0.33	0.29435×10^{-3}	0.108	0.758	108.910	2.880	3.85	C	15.2
403.	7649.	0.47	0.38298×10^{-3}	0.107	0.758	108.910	2.050	3.86	C	14.5
439.	7603.	0.68	0.50291×10^{-3}	0.105	0.756	108.910	1.440	3.86	C	13.9
297.	10593.	0.06	0.11992×10^{-3}	0.117	0.765	108.910	12.400	3.81	C	21.2
301.	10588.	0.10	0.13595×10^{-3}	0.116	0.765	108.910	8.870	3.81	C	20.0
318.	10565.	0.16	0.17605×10^{-3}	0.116	0.764	108.910	5.560	3.81	C	18.5
338.	10528.	0.24	0.22772×10^{-3}	0.115	0.763	108.910	3.820	3.82	C	17.7
360.	10473.	0.35	0.29563×10^{-3}	0.113	0.762	108.910	2.690	3.83	C	17.2
381.	10393.	0.51	0.38701×10^{-3}	0.111	0.760	108.910	1.890	3.84	C	17.1
414.	5020.	0.01	0.31866×10^{-3}	0.112	0.762	108.910	41.850	3.83	P	19.9
409.	5016.	0.14	0.25133×10^{-3}	0.112	0.761	108.910	6.570	3.83	P	23.5
320.	7707.	0.07	0.39148×10^{-3}	0.111	0.760	108.910	11.920	3.84	P	11.3
312.	7691.	0.23	0.30912×10^{-3}	0.110	0.760	108.910	4.130	3.84	P	12.3
306.	7669.	0.34	0.24302×10^{-3}	0.109	0.759	108.910	2.790	3.84	P	14.0
302.	7643.	0.43	0.19075×10^{-3}	0.108	0.759	108.910	2.260	3.85	P	16.6
305.	10264.	0.13	0.26885×10^{-3}	0.121	0.768	108.910	6.610	3.79	P	11.1
291.	10234.	0.23	0.20769×10^{-3}	0.121	0.767	108.910	3.970	3.79	P	12.5
280.	10198.	0.30	0.15987×10^{-3}	0.119	0.767	108.910	3.090	3.80	P	14.6
271.	10147.	0.37	0.11474×10^{-3}	0.118	0.766	108.910	2.590	3.81	P	18.6
270.	10114.	0.40	0.94450×10^{-4}	0.117	0.765	108.910	2.420	3.81	P	22.1
235.	5259.	0.11	0.12665×10^{-3}	0.108	0.758	108.910	7.670	3.85	C	24.8
267.	5255.	0.18	0.16133×10^{-3}	0.108	0.758	108.910	5.200	3.85	C	23.7
285.	5250.	0.25	0.19218×10^{-3}	0.107	0.758	108.910	3.800	3.85	C	22.9
292.	5243.	0.33	0.22260×10^{-3}	0.107	0.758	108.910	2.860	3.86	C	22.5
291.	5232.	0.44	0.25424×10^{-3}	0.106	0.757	108.910	2.180	3.86	C	22.3
249.	5325.	0.11	0.14638×10^{-3}	0.109	0.759	108.910	8.080	3.84	C	23.1

265.	5323.	0.15	0.16738×10^{-3}	0.109	0.759	108.910	6.130	3.84	C	22.2
291.	5318.	0.23	0.20977×10^{-3}	0.109	0.759	108.910	4.100	3.84	C	21.2
313.	5311.	0.32	0.25656×10^{-3}	0.109	0.759	108.910	2.960	3.85	C	20.6
334.	5301.	0.44	0.31239×10^{-3}	0.108	0.758	108.910	2.190	3.85	C	20.3
353.	5287.	0.60	0.38241×10^{-3}	0.107	0.758	108.910	1.630	3.85	C	20.2
292.	4916.	0.05	0.35484×10^{-3}	0.107	0.757	108.910	15.850	3.86	P	14.1
270.	4912.	0.19	0.27843×10^{-3}	0.107	0.757	108.910	4.860	3.86	P	15.2
251.	4906.	0.29	0.21838×10^{-3}	0.106	0.757	108.910	3.230	3.86	P	17.0
234.	4898.	0.37	0.17216×10^{-3}	0.106	0.757	108.910	2.600	3.86	P	19.3
219.	4888.	0.44	0.12958×10^{-3}	0.105	0.756	108.910	2.210	3.86	P	23.2
247.	5318.	0.11	0.15097×10^{-3}	0.109	0.759	108.910	7.860	3.84	C	22.2
266.	5315.	0.15	0.17612×10^{-3}	0.109	0.759	108.910	5.940	3.84	C	21.3
292.	5310.	0.24	0.22141×10^{-3}	0.109	0.759	108.910	3.940	3.85	C	20.2
310.	5302.	0.33	0.26611×10^{-3}	0.109	0.759	108.910	2.840	3.85	C	19.7
322.	5292.	0.46	0.31490×10^{-3}	0.108	0.758	108.910	2.120	3.85	C	19.5
330.	5279.	0.61	0.37147×10^{-3}	0.107	0.758	108.910	1.590	3.85	C	19.5
416.	5335.	0.13	0.29771×10^{-3}	0.119	0.766	108.910	6.860	3.80	P	20.3
411.	5329.	0.24	0.23342×10^{-3}	0.118	0.766	108.910	3.900	3.80	P	24.2

Convective Boiling of R1234ze(E) within a micro-fin tube

(file: GWPNU.dat)

Nu	Re	x_q	Bo	P_s/P_c	T_s/T_c	M_w	S_v	Pr	flow	U_{Nu}
244.	4334.	0.11	0.13621×10^{-3}	0.074	0.730	114.040	8.280	4.08	C	24.5
282.	4329.	0.18	0.17609×10^{-3}	0.074	0.730	114.040	5.250	4.08	C	23.1
318.	4322.	0.26	0.22403×10^{-3}	0.073	0.729	114.040	3.650	4.08	C	21.8
353.	4312.	0.37	0.28426×10^{-3}	0.073	0.729	114.040	2.600	4.09	C	20.7
387.	4297.	0.53	0.36244×10^{-3}	0.072	0.728	114.040	1.850	4.10	C	19.5
228.	6670.	0.06	0.11317×10^{-3}	0.075	0.731	114.040	13.060	4.07	C	20.6
235.	6666.	0.10	0.12667×10^{-3}	0.075	0.731	114.040	9.310	4.07	C	19.9
260.	6649.	0.16	0.16753×10^{-3}	0.074	0.730	114.040	5.790	4.08	C	18.5
291.	6622.	0.24	0.22503×10^{-3}	0.073	0.729	114.040	3.910	4.08	C	17.3
322.	6582.	0.36	0.30300×10^{-3}	0.072	0.728	114.040	2.680	4.10	C	16.2
352.	6523.	0.54	0.40843×10^{-3}	0.070	0.726	114.040	1.820	4.11	C	15.2
237.	9286.	0.03	0.10420×10^{-3}	0.082	0.738	114.040	20.770	4.01	C	23.7
230.	9282.	0.06	0.11341×10^{-3}	0.082	0.738	114.040	13.600	4.01	C	22.6
243.	9257.	0.11	0.15038×10^{-3}	0.082	0.737	114.040	7.870	4.02	C	20.8
270.	9211.	0.18	0.20752×10^{-3}	0.080	0.736	114.040	5.080	4.03	C	19.4
300.	9140.	0.29	0.28732×10^{-3}	0.079	0.734	114.040	3.340	4.04	C	18.2
330.	9034.	0.45	0.39585×10^{-3}	0.076	0.732	114.040	2.180	4.06	C	17.1
228.	4437.	0.01	0.25508×10^{-3}	0.072	0.728	114.040	49.960	4.10	P	16.2
223.	4432.	0.12	0.19937×10^{-3}	0.072	0.727	114.040	7.770	4.10	P	18.1
221.	4425.	0.20	0.15564×10^{-3}	0.071	0.727	114.040	4.810	4.10	P	21.1
284.	6960.	0.13	0.28961×10^{-3}	0.081	0.737	114.040	6.790	4.02	P	13.1
286.	6939.	0.24	0.22333×10^{-3}	0.081	0.736	114.040	4.000	4.02	P	15.4
294.	6914.	0.31	0.17058×10^{-3}	0.080	0.736	114.040	3.090	4.03	P	19.2
206.	6751.	0.01	0.23563×10^{-3}	0.077	0.733	114.040	56.430	4.05	P	10.3
206.	6739.	0.10	0.18371×10^{-3}	0.077	0.732	114.040	8.830	4.06	P	12.1
210.	6721.	0.17	0.14309×10^{-3}	0.076	0.732	114.040	5.460	4.06	P	14.7
221.	6699.	0.22	0.11216×10^{-3}	0.075	0.731	114.040	4.270	4.07	P	18.8
398.	4606.	0.40	0.30454×10^{-3}	0.073	0.729	114.040	2.450	4.09	C	23.4
427.	4587.	0.57	0.39570×10^{-3}	0.072	0.728	114.040	1.730	4.10	C	21.8
206.	6333.	0.08	0.10391×10^{-3}	0.070	0.725	114.040	11.310	4.12	C	20.0
229.	6330.	0.11	0.12313×10^{-3}	0.069	0.725	114.040	8.380	4.12	C	19.6
262.	6313.	0.18	0.16604×10^{-3}	0.069	0.725	114.040	5.340	4.12	C	18.6
285.	6284.	0.26	0.21675×10^{-3}	0.068	0.724	114.040	3.670	4.13	C	17.4
299.	6241.	0.38	0.27922×10^{-3}	0.067	0.722	114.040	2.570	4.15	C	16.3
307.	6177.	0.55	0.35829×10^{-3}	0.065	0.720	114.040	1.800	4.17	C	15.1
216.	9439.	0.05	0.10544×10^{-3}	0.086	0.741	114.040	16.790	3.98	C	19.6
223.	9435.	0.07	0.11866×10^{-3}	0.086	0.741	114.040	11.880	3.99	C	18.9
252.	9410.	0.12	0.16026×10^{-3}	0.085	0.740	114.040	7.260	3.99	C	17.7
285.	9367.	0.19	0.21893×10^{-3}	0.084	0.739	114.040	4.820	4.00	C	16.6
319.	9300.	0.30	0.29730×10^{-3}	0.082	0.738	114.040	3.240	4.01	C	15.6

349.	9200.	0.45	0.40073×10^{-3}	0.080	0.736	114.040	2.150	4.03	C	14.7
239.	4661.	0.01	0.30656×10^{-3}	0.078	0.734	114.040	50.880	4.05	P	12.4
234.	4656.	0.13	0.24249×10^{-3}	0.077	0.733	114.040	7.000	4.05	P	14.1
229.	4648.	0.22	0.19056×10^{-3}	0.077	0.733	114.040	4.280	4.05	P	16.7
226.	4638.	0.29	0.14916×10^{-3}	0.077	0.733	114.040	3.340	4.06	P	20.2
253.	7202.	0.10	0.30516×10^{-3}	0.086	0.741	114.040	8.530	3.98	P	11.0
249.	7180.	0.20	0.23161×10^{-3}	0.085	0.741	114.040	4.620	3.99	P	13.0
250.	7153.	0.27	0.17599×10^{-3}	0.085	0.740	114.040	3.480	3.99	P	16.0
265.	7113.	0.34	0.12624×10^{-3}	0.083	0.739	114.040	2.850	4.00	P	22.2
207.	6627.	0.09	0.18755×10^{-3}	0.077	0.733	114.040	9.930	4.06	P	12.1
208.	6611.	0.16	0.14641×10^{-3}	0.076	0.732	114.040	5.820	4.06	P	14.6
214.	6590.	0.21	0.11455×10^{-3}	0.075	0.731	114.040	4.460	4.07	P	18.1
205.	4261.	0.07	0.11900×10^{-3}	0.070	0.726	114.040	12.220	4.12	C	24.3
220.	4259.	0.11	0.13760×10^{-3}	0.070	0.726	114.040	8.580	4.12	C	23.4
246.	4253.	0.18	0.17764×10^{-3}	0.070	0.725	114.040	5.270	4.12	C	22.1
268.	4245.	0.27	0.22404×10^{-3}	0.069	0.725	114.040	3.610	4.12	C	21.4
289.	4233.	0.38	0.28098×10^{-3}	0.069	0.724	114.040	2.550	4.13	C	20.9
309.	4217.	0.54	0.35357×10^{-3}	0.068	0.723	114.040	1.810	4.14	C	20.7
238.	6791.	0.06	0.10646×10^{-3}	0.074	0.730	114.040	13.810	4.08	C	21.9
245.	6787.	0.09	0.11875×10^{-3}	0.074	0.730	114.040	9.830	4.08	C	21.3
269.	6771.	0.15	0.15457×10^{-3}	0.073	0.729	114.040	6.130	4.08	C	19.8
295.	6745.	0.23	0.20411×10^{-3}	0.072	0.728	114.040	4.180	4.09	C	18.5
320.	6706.	0.33	0.27080×10^{-3}	0.071	0.727	114.040	2.890	4.10	C	17.2
343.	6648.	0.49	0.36063×10^{-3}	0.069	0.725	114.040	1.990	4.12	C	16.1
217.	9402.	0.04	0.95610×10^{-4}	0.081	0.736	114.040	19.080	4.02	C	22.6
216.	9398.	0.06	0.10598×10^{-3}	0.081	0.736	114.040	13.070	4.02	C	21.8
233.	9372.	0.12	0.14118×10^{-3}	0.080	0.736	114.040	7.810	4.03	C	20.3
258.	9326.	0.18	0.19264×10^{-3}	0.079	0.735	114.040	5.140	4.04	C	19.1
285.	9256.	0.28	0.26298×10^{-3}	0.077	0.733	114.040	3.430	4.05	C	18.1
310.	9151.	0.43	0.35762×10^{-3}	0.075	0.731	114.040	2.280	4.07	C	17.1
267.	2192.	0.06	0.34857×10^{-3}	0.071	0.726	114.040	15.060	4.11	P	22.3
255.	2191.	0.21	0.27768×10^{-3}	0.070	0.726	114.040	4.590	4.11	P	25.0
248.	4364.	0.13	0.34036×10^{-3}	0.074	0.730	114.040	6.890	4.08	P	12.6
244.	4355.	0.27	0.26563×10^{-3}	0.074	0.730	114.040	3.620	4.08	P	14.6
242.	4342.	0.36	0.20601×10^{-3}	0.073	0.729	114.040	2.690	4.08	P	17.5
242.	4324.	0.44	0.14822×10^{-3}	0.072	0.728	114.040	2.200	4.09	P	23.3
194.	6492.	0.09	0.18181×10^{-3}	0.072	0.728	114.040	9.570	4.09	P	11.1
197.	6477.	0.17	0.14472×10^{-3}	0.072	0.728	114.040	5.640	4.10	P	13.2
204.	6460.	0.22	0.11488×10^{-3}	0.071	0.727	114.040	4.320	4.10	P	16.4
223.	6434.	0.27	0.85580×10^{-4}	0.070	0.726	114.040	3.560	4.11	P	23.2
275.	6786.	0.00	0.41468×10^{-3}	0.081	0.737	114.040	57.240	4.02	P	10.8
264.	6767.	0.16	0.32230×10^{-3}	0.081	0.736	114.040	5.670	4.02	P	11.6
254.	6744.	0.28	0.24723×10^{-3}	0.080	0.736	114.040	3.430	4.03	P	13.1
244.	6719.	0.36	0.18713×10^{-3}	0.079	0.735	114.040	2.670	4.04	P	15.3
233.	6685.	0.43	0.12843×10^{-3}	0.078	0.734	114.040	2.250	4.05	P	20.0
229.	6664.	0.46	0.10087×10^{-3}	0.077	0.733	114.040	2.110	4.05	P	24.5
323.	4464.	0.26	0.21255×10^{-3}	0.072	0.728	114.040	3.680	4.09	C	24.1
345.	4454.	0.36	0.26266×10^{-3}	0.072	0.728	114.040	2.660	4.10	C	23.0
363.	4438.	0.51	0.32560×10^{-3}	0.071	0.727	114.040	1.920	4.10	C	21.9

222.	6590.	0.06	0.11322×10^{-3}	0.074	0.730	114.040	13.420	4.08	C	20.2
230.	6586.	0.09	0.12763×10^{-3}	0.074	0.730	114.040	9.450	4.08	C	19.5
250.	6570.	0.16	0.16672×10^{-3}	0.073	0.729	114.040	5.810	4.09	C	18.1
273.	6544.	0.24	0.21889×10^{-3}	0.072	0.728	114.040	3.930	4.09	C	16.9
295.	6505.	0.36	0.28783×10^{-3}	0.071	0.727	114.040	2.720	4.10	C	15.9
315.	6448.	0.52	0.37957×10^{-3}	0.069	0.725	114.040	1.870	4.12	C	15.0
213.	9097.	0.05	0.10171×10^{-3}	0.081	0.737	114.040	16.790	4.02	C	21.0
214.	9092.	0.07	0.11344×10^{-3}	0.081	0.737	114.040	11.750	4.02	C	20.2
230.	9067.	0.13	0.14996×10^{-3}	0.080	0.736	114.040	7.160	4.03	C	18.8
252.	9022.	0.20	0.20148×10^{-3}	0.079	0.735	114.040	4.780	4.03	C	17.7
275.	8954.	0.30	0.27074×10^{-3}	0.078	0.733	114.040	3.240	4.05	C	16.8
298.	8854.	0.45	0.36299×10^{-3}	0.075	0.731	114.040	2.180	4.07	C	16.0
263.	2205.	0.05	0.32156×10^{-3}	0.070	0.726	114.040	16.970	4.12	P	23.2
274.	4444.	0.00	0.31468×10^{-3}	0.075	0.730	114.040	61.700	4.07	P	14.6
277.	4439.	0.14	0.25573×10^{-3}	0.074	0.730	114.040	6.700	4.08	P	16.6
279.	4431.	0.24	0.20500×10^{-3}	0.074	0.730	114.040	4.010	4.08	P	19.6
279.	4421.	0.31	0.16170×10^{-3}	0.073	0.729	114.040	3.090	4.08	P	23.7
277.	4606.	0.17	0.33844×10^{-3}	0.078	0.734	114.040	5.460	4.05	P	14.1
259.	4596.	0.30	0.26302×10^{-3}	0.077	0.733	114.040	3.240	4.05	P	15.7
238.	4585.	0.39	0.19975×10^{-3}	0.077	0.733	114.040	2.510	4.06	P	17.9
205.	4569.	0.46	0.13410×10^{-3}	0.076	0.732	114.040	2.110	4.06	P	21.8
180.	4559.	0.49	0.10125×10^{-3}	0.076	0.731	114.040	1.980	4.07	P	25.0
192.	6406.	0.07	0.17881×10^{-3}	0.073	0.729	114.040	12.570	4.09	P	11.8
200.	6390.	0.14	0.14326×10^{-3}	0.072	0.728	114.040	6.610	4.09	P	14.2
210.	6370.	0.19	0.11382×10^{-3}	0.072	0.727	114.040	4.880	4.10	P	17.7
292.	6768.	0.04	0.39651×10^{-3}	0.082	0.737	114.040	19.670	4.02	P	11.4
290.	6750.	0.19	0.31129×10^{-3}	0.081	0.737	114.040	4.950	4.02	P	12.6
289.	6728.	0.30	0.24141×10^{-3}	0.080	0.736	114.040	3.190	4.03	P	14.7
292.	6702.	0.38	0.18494×10^{-3}	0.080	0.735	114.040	2.540	4.03	P	18.0
286.	2202.	0.02	0.39490×10^{-3}	0.077	0.733	114.040	34.690	4.06	P	23.3
236.	4348.	0.02	0.26964×10^{-3}	0.071	0.726	114.040	29.860	4.11	P	13.5
238.	4343.	0.14	0.21689×10^{-3}	0.070	0.726	114.040	6.610	4.11	P	15.7
240.	4335.	0.23	0.17294×10^{-3}	0.070	0.726	114.040	4.180	4.11	P	18.7
243.	4325.	0.29	0.13673×10^{-3}	0.070	0.725	114.040	3.280	4.12	P	23.1
259.	4431.	0.05	0.36018×10^{-3}	0.073	0.729	114.040	17.450	4.09	P	12.9
253.	4424.	0.20	0.28945×10^{-3}	0.072	0.728	114.040	4.720	4.09	P	14.1
245.	4415.	0.32	0.22889×10^{-3}	0.072	0.728	114.040	3.060	4.10	P	15.9
233.	4403.	0.40	0.17746×10^{-3}	0.071	0.727	114.040	2.440	4.10	P	18.4
210.	4387.	0.47	0.12333×10^{-3}	0.071	0.727	114.040	2.070	4.11	P	23.0

APPENDIX B1

Convective Boiling of R134a within a micro-fin tube

(file: lowgwpq.dat)

q'' (Wm^{-2})	ΔT_s (K)	x_q	G_r (kg m^{-2}s^{-1})	T_w (K)	T_s (K)	T_f (K)	z (m)	M_w (g/mole)	flow
15040.	3.84	0.11	326.	285.50	281.70	292.20	0.97	102.03	P
11240.	3.18	0.19	326.	284.70	281.50	289.80	1.54	102.03	P
8476.	2.62	0.26	326.	283.90	281.30	287.80	2.06	102.03	P
6153.	2.05	0.31	326.	283.00	281.00	286.30	2.66	102.03	P
5270.	1.77	0.33	326.	282.50	280.70	285.20	3.00	102.03	P
14260.	3.73	0.09	332.	285.00	281.30	291.40	0.97	102.03	P
10710.	3.09	0.17	332.	284.20	281.10	289.10	1.54	102.03	P
8110.	2.54	0.23	332.	283.50	280.90	287.20	2.06	102.03	P
5900.	1.97	0.28	332.	282.60	280.60	285.70	2.66	102.03	P
5046.	1.69	0.30	332.	282.10	280.40	284.70	3.00	102.03	P
13380.	3.66	0.09	396.	284.80	281.10	290.60	0.97	102.03	P
10080.	3.02	0.16	396.	284.00	281.00	288.50	1.54	102.03	P
7650.	2.49	0.20	396.	283.20	280.80	286.70	2.06	102.03	P
5555.	1.93	0.24	396.	282.40	280.50	285.30	2.66	102.03	P
4728.	1.65	0.26	396.	281.90	280.30	284.40	3.00	102.03	P
13290.	3.64	0.09	400.	284.70	281.00	290.40	0.97	102.03	P
10030.	3.01	0.15	400.	283.90	280.80	288.40	1.54	102.03	P
7609.	2.47	0.19	399.	283.10	280.60	286.60	2.06	102.03	P
5520.	1.92	0.23	399.	282.30	280.30	285.20	2.66	102.03	P
4689.	1.64	0.25	399.	281.80	280.10	284.20	3.00	102.03	P
8400.	1.73	0.33	198.	280.50	278.70	286.50	5.72	102.03	C
9846.	1.97	0.44	198.	280.50	278.60	288.20	6.34	102.03	C
2620.	1.39	0.03	308.	280.70	279.30	282.00	2.66	102.03	C
2960.	1.52	0.04	308.	280.80	279.20	282.50	3.00	102.03	C
6318.	1.92	0.08	308.	280.90	279.00	283.00	3.69	102.03	C
7140.	2.03	0.11	308.	281.00	279.00	284.20	4.03	102.03	C
9165.	2.31	0.17	308.	281.20	278.90	286.30	4.61	102.03	C
11760.	2.65	0.24	308.	281.30	278.70	288.00	5.15	102.03	C
15160.	3.09	0.33	308.	281.50	278.40	290.80	5.72	102.03	C
19740.	3.69	0.46	308.	281.60	277.90	294.40	6.34	102.03	C
5787.	2.07	0.04	414.	281.30	279.20	283.20	3.69	102.03	C
6675.	2.25	0.06	414.	281.40	279.20	284.30	4.03	102.03	C
8668.	2.63	0.10	414.	281.70	279.10	286.30	4.61	102.03	C
11060.	3.08	0.15	414.	281.90	278.80	287.90	5.15	102.03	C
14090.	3.65	0.22	414.	282.20	278.50	290.50	5.72	102.03	C
18060.	4.38	0.31	414.	282.40	278.10	293.80	6.34	102.03	C
17980.	4.26	0.09	199.	283.30	279.00	292.00	0.35	102.03	P
13960.	3.81	0.28	199.	282.70	278.80	288.90	0.97	102.03	P
10740.	3.41	0.41	199.	282.10	278.70	286.70	1.54	102.03	P
13650.	3.64	0.09	306.	283.50	279.90	289.50	0.97	102.03	P
10570.	3.01	0.18	306.	282.70	279.70	287.40	1.54	102.03	P

8229.	2.49	0.24	306.	282.00	279.50	285.50	2.06	102.03	P
6107.	1.96	0.30	306.	281.10	279.20	284.10	2.66	102.03	P
5203.	1.70	0.32	306.	280.70	279.00	283.20	3.00	102.03	P
12730.	3.53	0.08	407.	284.20	280.70	289.60	0.97	102.03	P
9829.	2.90	0.14	407.	283.40	280.50	287.70	1.54	102.03	P
7630.	2.38	0.18	407.	282.70	280.30	286.00	2.06	102.03	P
5652.	1.84	0.22	407.	281.90	280.00	284.70	2.66	102.03	P
4818.	1.57	0.24	407.	281.40	279.80	283.80	3.00	102.03	P
2959.	1.41	0.04	226.	281.80	280.40	283.50	3.00	102.03	C
6249.	1.73	0.10	226.	282.00	280.20	284.00	3.69	102.03	C
6988.	1.84	0.14	226.	282.10	280.20	285.10	4.03	102.03	C
8792.	2.08	0.21	226.	282.20	280.10	287.10	4.61	102.03	C
11090.	2.34	0.30	226.	282.30	280.00	288.80	5.15	102.03	C
14090.	2.68	0.42	226.	282.50	279.80	291.40	5.72	102.03	C
18120.	3.10	0.59	226.	282.60	279.50	294.70	6.34	102.03	C
2966.	1.53	0.02	303.	282.10	280.50	283.40	2.66	102.03	C
3462.	1.70	0.03	303.	282.20	280.50	284.10	3.00	102.03	C
8025.	2.19	0.08	303.	282.50	280.30	284.80	3.69	102.03	C
8727.	2.36	0.12	303.	282.60	280.20	286.20	4.03	102.03	C
11100.	2.73	0.19	303.	282.80	280.10	289.10	4.61	102.03	C
14660.	3.17	0.28	303.	283.00	279.90	291.50	5.15	102.03	C
19700.	3.73	0.40	303.	283.30	279.50	295.00	5.72	102.03	C
26840.	4.48	0.58	303.	283.50	279.00	300.20	6.34	102.03	C
3309.	1.83	0.01	407.	284.10	282.30	285.80	3.00	102.03	C
8453.	2.51	0.05	407.	284.50	282.00	286.60	3.69	102.03	C
9097.	2.75	0.08	407.	284.70	282.00	288.10	4.03	102.03	C
11650.	3.28	0.13	407.	285.10	281.80	291.20	4.61	102.03	C
15680.	3.90	0.20	407.	285.40	281.50	293.90	5.15	102.03	C
21530.	4.68	0.30	407.	285.80	281.10	297.90	5.72	102.03	C
29940.	5.70	0.46	407.	286.20	280.50	303.90	6.34	102.03	C
17690.	4.16	0.00	209.	283.40	279.20	292.10	0.35	102.03	P
13960.	3.50	0.18	209.	282.60	279.10	289.00	0.97	102.03	P
11010.	2.93	0.31	209.	281.90	279.00	286.90	1.54	102.03	P
8705.	2.45	0.40	209.	281.30	278.90	285.00	2.06	102.03	P
6545.	1.94	0.49	209.	280.60	278.60	283.50	2.66	102.03	P
18340.	4.45	0.09	306.	287.50	283.00	295.10	0.97	102.03	P
13900.	3.68	0.21	306.	286.60	282.90	292.30	1.54	102.03	P
10610.	3.04	0.29	306.	285.70	282.60	289.80	2.06	102.03	P
7783.	2.36	0.36	306.	284.70	282.30	288.10	2.66	102.03	P
6666.	2.02	0.39	306.	284.10	282.10	286.80	3.00	102.03	P
12290.	3.41	0.06	405.	284.50	281.10	289.60	0.97	102.03	P
9516.	2.81	0.12	405.	283.70	280.90	287.80	1.54	102.03	P
7427.	2.29	0.16	405.	283.00	280.70	286.10	2.06	102.03	P
5569.	1.77	0.20	405.	282.20	280.40	284.90	2.66	102.03	P
4798.	1.50	0.22	405.	281.70	280.20	284.00	3.00	102.03	P
2815.	1.39	0.04	219.	280.80	279.40	282.10	2.66	102.03	C
3309.	1.50	0.06	219.	280.80	279.30	282.70	3.00	102.03	C
6646.	1.84	0.12	219.	281.00	279.20	283.30	3.69	102.03	C
7665.	1.95	0.16	219.	281.10	279.10	284.50	4.03	102.03	C

9768.	2.20	0.25	219.	281.20	279.00	286.80	4.61	102.03	C
12130.	2.47	0.35	219.	281.40	278.90	288.70	5.15	102.03	C
15000.	2.80	0.48	219.	281.50	278.70	291.50	5.72	102.03	C
18630.	3.23	0.66	219.	281.60	278.40	295.00	6.34	102.03	C
3210.	1.67	0.02	314.	282.00	280.30	284.60	3.00	102.03	C
8631.	2.36	0.07	314.	282.50	280.20	285.70	3.69	102.03	C
9642.	2.64	0.11	314.	282.80	280.20	288.10	4.03	102.03	C
12990.	3.23	0.19	314.	283.20	280.00	293.20	4.61	102.03	C
17960.	3.91	0.29	314.	283.70	279.70	297.30	5.15	102.03	C
24970.	4.74	0.44	314.	284.10	279.30	304.30	5.72	102.03	C
34900.	5.82	0.66	314.	284.50	278.70	314.60	6.34	102.03	C
9654.	2.45	0.08	319.	283.40	281.00	286.50	3.69	102.03	C
10790.	2.75	0.12	319.	283.70	281.00	288.90	4.03	102.03	C
14530.	3.38	0.21	319.	284.20	280.80	294.30	4.61	102.03	C
20060.	4.08	0.32	319.	284.60	280.50	298.70	5.15	102.03	C
27880.	4.94	0.48	319.	285.00	280.10	306.00	5.72	102.03	C
38920.	6.04	0.73	319.	285.50	279.50	316.80	6.34	102.03	C
14530.	3.85	0.14	404.	290.10	286.20	298.30	4.61	102.03	C
20590.	4.89	0.23	404.	290.80	286.00	302.80	5.15	102.03	C
29420.	6.09	0.37	404.	291.70	285.60	309.80	5.72	102.03	C
42170.	7.56	0.58	404.	292.60	285.00	320.80	6.34	102.03	C
19100.	4.18	0.09	211.	283.20	279.00	293.50	0.35	102.03	P
14850.	3.62	0.28	210.	282.50	278.90	289.90	0.97	102.03	P
11440.	3.16	0.42	210.	281.90	278.70	287.30	1.54	102.03	P
29830.	5.84	0.08	310.	290.30	284.50	303.10	0.35	102.03	P
22260.	4.94	0.28	310.	289.30	284.30	297.90	0.97	102.03	P
16500.	4.15	0.41	310.	288.30	284.10	294.50	1.54	102.03	P

APPENDIX B2

Convective Boiling of R1234yf/ R134a (56/44) within a micro-fin tube

(file: lowgwpq.dat)

q'' (Wm^{-2})	ΔT_s (K)	x_q	G_r (kg m^{-2}s^{-1})	T_w (K)	T_s (K)	T_f (K)	z (m)	M_w (g/mole)	flow
5823.	1.49	0.18	208.	279.20	277.70	282.90	4.61	108.91	C
6919.	1.66	0.25	208.	279.20	277.60	284.00	5.15	108.91	C
8079.	1.88	0.33	208.	279.30	277.50	285.50	5.72	108.91	C
9382.	2.15	0.44	208.	279.50	277.30	287.20	6.34	108.91	C
3041.	1.44	0.03	228.	280.20	278.70	282.00	3.00	108.91	C
6629.	1.83	0.10	228.	280.40	278.60	282.60	3.69	108.91	C
7431.	1.99	0.14	228.	280.50	278.50	283.80	4.03	108.91	C
9253.	2.29	0.23	228.	280.70	278.40	286.00	4.61	108.91	C
11460.	2.62	0.34	228.	280.90	278.30	287.80	5.15	108.91	C
14270.	3.00	0.47	228.	281.10	278.10	290.50	5.72	108.91	C
17960.	3.48	0.66	228.	281.30	277.90	293.90	6.34	108.91	C
6228.	1.70	0.09	308.	281.70	279.90	283.80	3.69	108.91	C
6990.	1.83	0.12	308.	281.70	279.90	285.00	4.03	108.91	C
8845.	2.11	0.19	308.	281.90	279.80	287.10	4.61	108.91	C
11200.	2.43	0.26	308.	282.00	279.60	289.00	5.15	108.91	C
14280.	2.81	0.36	308.	282.20	279.40	291.70	5.72	108.91	C
18410.	3.30	0.50	308.	282.40	279.10	295.20	6.34	108.91	C
3313.	1.76	0.03	304.	280.10	278.40	282.40	2.66	108.91	C
3782.	1.98	0.05	304.	280.30	278.30	283.30	3.00	108.91	C
9843.	2.64	0.11	304.	280.60	278.00	284.20	3.69	108.91	C
11990.	2.89	0.16	304.	280.80	277.90	286.40	4.03	108.91	C
15860.	3.38	0.28	304.	281.10	277.70	290.60	4.61	108.91	C
19670.	3.93	0.41	304.	281.40	277.40	294.40	5.15	108.91	C
23840.	4.60	0.59	304.	281.60	277.00	300.00	5.72	108.91	C
28680.	5.45	0.82	304.	281.90	276.50	306.50	6.34	108.91	C
2730.	1.59	0.01	414.	282.10	280.50	284.10	2.66	108.91	C
3083.	1.82	0.02	414.	282.20	280.40	284.70	3.00	108.91	C
8165.	2.52	0.07	414.	282.60	280.10	285.40	3.69	108.91	C
9160.	2.76	0.10	414.	282.80	280.10	287.10	4.03	108.91	C
12000.	3.28	0.16	414.	283.20	279.90	290.30	4.61	108.91	C
15930.	3.88	0.24	414.	283.50	279.60	293.10	5.15	108.91	C
21310.	4.64	0.35	414.	283.80	279.10	297.20	5.72	108.91	C
28760.	5.63	0.51	414.	284.10	278.50	303.20	6.34	108.91	C
9441.	2.35	0.04	203.	282.90	280.50	288.10	0.35	108.91	P
7528.	2.03	0.15	203.	282.50	280.50	286.40	0.97	108.91	P
5944.	1.75	0.23	203.	282.20	280.40	285.20	1.54	108.91	P
4642.	1.51	0.29	203.	281.80	280.30	284.10	2.06	108.91	P
14660.	3.49	0.19	221.	283.30	279.80	290.20	0.97	108.91	P
11320.	2.94	0.33	221.	282.60	279.70	287.90	1.54	108.91	P
8763.	2.46	0.43	221.	282.00	279.60	285.80	2.06	108.91	P
16980.	3.88	0.17	299.	284.50	280.60	291.90	0.97	108.91	P

12960.	3.18	0.29	298.	283.60	280.50	289.30	1.54	108.91	P
9914.	2.60	0.38	298.	282.80	280.20	286.90	2.06	108.91	P
7189.	1.98	0.45	298.	281.90	279.90	285.20	2.66	108.91	P
6046.	1.67	0.48	299.	281.40	279.70	284.10	3.00	108.91	P
16930.	4.04	0.09	397.	284.90	280.90	292.20	0.97	108.91	P
12900.	3.26	0.18	397.	283.90	280.70	289.60	1.54	108.91	P
9852.	2.61	0.25	397.	283.00	280.40	287.20	2.06	108.91	P
7142.	1.93	0.30	397.	282.00	280.00	285.60	2.66	108.91	P
6016.	1.59	0.33	397.	281.40	279.80	284.40	3.00	108.91	P
5619.	1.68	0.09	216.	280.10	278.40	282.20	3.69	108.91	C
6449.	1.84	0.12	215.	280.20	278.40	283.40	4.03	108.91	C
8096.	2.15	0.21	215.	280.40	278.30	285.20	4.61	108.91	C
9890.	2.49	0.31	215.	280.70	278.20	286.70	5.15	108.91	C
12010.	2.87	0.43	216.	280.90	278.00	289.00	5.72	108.91	C
14650.	3.34	0.59	216.	281.10	277.80	291.80	6.34	108.91	C
3086.	1.49	0.04	303.	280.70	279.20	282.60	2.66	108.91	C
3510.	1.68	0.05	303.	280.90	279.20	283.30	3.00	108.91	C
8049.	2.22	0.11	303.	281.20	278.90	284.00	3.69	108.91	C
9259.	2.40	0.15	303.	281.30	278.90	285.60	4.03	108.91	C
12020.	2.80	0.24	303.	281.60	278.70	288.50	4.61	108.91	C
15380.	3.27	0.34	303.	281.80	278.50	291.00	5.15	108.91	C
19660.	3.86	0.48	303.	282.00	278.20	295.10	5.72	108.91	C
25300.	4.63	0.68	303.	282.30	277.70	300.10	6.34	108.91	C
8219.	2.47	0.06	406.	281.50	279.10	284.30	3.69	108.91	C
9361.	2.77	0.09	406.	281.80	279.00	286.00	4.03	108.91	C
12280.	3.40	0.16	406.	282.20	278.90	289.00	4.61	108.91	C
16090.	4.12	0.24	406.	282.70	278.50	291.70	5.15	108.91	C
21150.	5.00	0.35	406.	283.10	278.10	296.00	5.72	108.91	C
28020.	6.15	0.51	406.	283.60	277.40	301.80	6.34	108.91	C
2616.	1.55	0.02	413.	281.50	279.90	283.50	2.66	108.91	C
3080.	1.82	0.03	413.	281.70	279.90	284.40	3.00	108.91	C
8833.	2.59	0.07	413.	282.10	279.60	285.50	3.69	108.91	C
9876.	2.85	0.11	413.	282.40	279.50	288.00	4.03	108.91	C
13410.	3.43	0.18	413.	282.70	279.30	293.30	4.61	108.91	C
18680.	4.13	0.27	413.	283.10	279.00	297.40	5.15	108.91	C
26160.	5.01	0.40	413.	283.50	278.50	304.80	5.72	108.91	C
36760.	6.18	0.60	413.	283.90	277.70	315.60	6.34	108.91	C
12940.	3.01	0.04	205.	282.40	279.40	289.40	0.35	108.91	P
10190.	2.57	0.18	205.	281.90	279.30	287.10	0.97	108.91	P
7991.	2.20	0.29	205.	281.40	279.20	285.40	1.54	108.91	P
6266.	1.89	0.37	205.	281.00	279.10	284.00	2.06	108.91	P
4627.	1.56	0.44	205.	280.50	278.90	282.90	2.66	108.91	P
3885.	1.40	0.47	205.	280.20	278.80	282.20	3.00	108.91	P
16930.	4.04	0.09	397.	284.90	280.90	292.20	0.97	108.91	P
12900.	3.26	0.18	397.	283.90	280.70	289.60	1.54	108.91	P
9852.	2.61	0.25	397.	283.00	280.40	287.20	2.06	108.91	P
7142.	1.93	0.30	397.	282.00	280.00	285.60	2.66	108.91	P
6016.	1.59	0.33	397.	281.40	279.80	284.40	3.00	108.91	P
18400.	4.74	0.14	399.	286.00	281.30	294.10	0.97	108.91	P

14110.	3.84	0.24	399.	284.90	281.00	291.10	1.54	108.91	P
10870.	3.08	0.31	399.	283.80	280.70	288.50	2.06	108.91	P
7984.	2.28	0.37	399.	282.60	280.30	286.70	2.66	108.91	P
6780.	1.88	0.40	399.	281.90	280.10	285.40	3.00	108.91	P
5868.	1.58	0.11	214.	280.70	279.10	282.90	3.69	108.91	C
6680.	1.71	0.15	214.	280.80	279.10	284.00	4.03	108.91	C
8409.	1.95	0.24	214.	281.00	279.00	285.90	4.61	108.91	C
10400.	2.21	0.34	214.	281.10	278.90	287.60	5.15	108.91	C
12860.	2.52	0.47	214.	281.30	278.70	290.00	5.72	108.91	C
16020.	2.91	0.65	214.	281.40	278.50	293.00	6.34	108.91	C
3523.	1.54	0.04	301.	281.90	280.40	284.10	3.00	108.91	C
7846.	2.03	0.10	301.	282.20	280.20	284.80	3.69	108.91	C
8720.	2.21	0.14	301.	282.30	280.10	286.30	4.03	108.91	C
11150.	2.58	0.23	301.	282.60	280.00	289.10	4.61	108.91	C
14460.	3.01	0.32	301.	282.80	279.80	291.50	5.15	108.91	C
18970.	3.54	0.46	301.	283.10	279.50	295.10	5.72	108.91	C
25200.	4.23	0.65	301.	283.30	279.10	300.10	6.34	108.91	C
3247.	1.75	0.03	398.	281.20	279.50	283.60	3.00	108.91	C
8115.	2.46	0.07	397.	281.70	279.20	284.30	3.69	108.91	C
9146.	2.71	0.10	397.	281.90	279.20	286.00	4.03	108.91	C
11930.	3.25	0.17	397.	282.20	279.00	289.00	4.61	108.91	C
15670.	3.88	0.25	398.	282.60	278.70	291.60	5.15	108.91	C
20720.	4.66	0.36	397.	282.90	278.30	295.70	5.72	108.91	C
27650.	5.68	0.53	398.	283.30	277.60	301.20	6.34	108.91	C
7496.	2.03	0.09	212.	281.00	279.00	285.00	0.97	108.91	P
5926.	1.75	0.17	212.	280.60	278.90	283.80	1.54	108.91	P
4629.	1.52	0.22	212.	280.30	278.80	282.70	2.06	108.91	P
10290.	2.74	0.14	210.	281.20	278.40	286.50	0.97	108.91	P
8071.	2.33	0.25	210.	280.70	278.30	284.80	1.54	108.91	P
6318.	1.99	0.33	210.	280.20	278.20	283.30	2.06	108.91	P
4641.	1.64	0.39	210.	279.70	278.00	282.20	2.66	108.91	P
3875.	1.46	0.42	210.	279.40	277.90	281.50	3.00	108.91	P
20040.	4.89	0.05	299.	283.10	278.20	293.40	0.35	108.91	P
15770.	4.02	0.20	299.	282.10	278.00	289.80	0.97	108.91	P
12360.	3.29	0.32	299.	281.10	277.80	287.20	1.54	108.91	P
9698.	2.67	0.40	299.	280.20	277.60	284.90	2.06	108.91	P
7178.	2.04	0.48	299.	279.20	277.20	283.20	2.66	108.91	P
17190.	4.05	0.08	412.	284.90	280.90	292.60	0.97	108.91	P
13210.	3.25	0.17	412.	283.90	280.60	289.80	1.54	108.91	P
10150.	2.58	0.23	412.	283.00	280.40	287.40	2.06	108.91	P
7336.	1.89	0.29	412.	281.90	280.00	285.70	2.66	108.91	P
6111.	1.54	0.32	412.	281.30	279.70	284.60	3.00	108.91	P
11870.	2.18	0.55	212.	279.70	277.60	289.70	6.34	108.91	C
8220.	1.91	0.11	303.	279.60	277.70	282.60	3.69	108.91	C
9419.	2.08	0.15	303.	279.70	277.70	284.20	4.03	108.91	C
12090.	2.47	0.24	303.	280.00	277.50	287.20	4.61	108.91	C
15270.	2.92	0.34	303.	280.20	277.20	289.70	5.15	108.91	C
19270.	3.50	0.48	303.	280.40	276.90	293.60	5.72	108.91	C
24510.	4.26	0.67	303.	280.60	276.40	298.60	6.34	108.91	C

8495.	2.17	0.07	390.	282.50	280.30	285.30	3.69	108.91	C
9570.	2.40	0.11	390.	282.70	280.30	287.00	4.03	108.91	C
12440.	2.90	0.18	390.	283.00	280.10	290.20	4.61	108.91	C
16270.	3.49	0.27	390.	283.30	279.80	293.00	5.15	108.91	C
21430.	4.23	0.38	390.	283.60	279.40	297.50	5.72	108.91	C
28500.	5.20	0.56	390.	284.00	278.80	303.50	6.34	108.91	C
7331.	1.59	0.07	205.	281.60	280.00	285.40	0.97	108.91	P
5733.	1.34	0.15	205.	281.30	280.00	284.20	1.54	108.91	P
21540.	4.74	0.03	306.	283.10	278.40	294.10	0.35	108.91	P
16850.	3.83	0.20	306.	282.00	278.20	290.20	0.97	108.91	P
13130.	3.05	0.31	306.	281.00	278.00	287.40	1.54	108.91	P
10230.	2.41	0.40	306.	280.10	277.70	285.00	2.06	108.91	P
7491.	1.74	0.48	306.	279.00	277.30	283.20	2.66	108.91	P
16350.	3.56	0.10	401.	285.60	282.10	293.30	0.97	108.91	P
12710.	2.81	0.18	401.	284.70	281.90	290.50	1.54	108.91	P
9818.	2.18	0.25	401.	283.80	281.60	288.10	2.06	108.91	P
7018.	1.51	0.31	401.	282.80	281.30	286.40	2.66	108.91	P
10810.	1.85	0.43	201.	281.30	279.50	288.60	5.72	108.91	C
13610.	2.17	0.59	201.	281.50	279.30	291.20	6.34	108.91	C
3386.	1.45	0.04	304.	280.10	278.70	282.70	3.00	108.91	C
8215.	2.01	0.10	304.	280.40	278.40	283.40	3.69	108.91	C
9276.	2.20	0.14	304.	280.50	278.30	285.10	4.03	108.91	C
11990.	2.60	0.23	304.	280.80	278.20	288.10	4.61	108.91	C
15530.	3.07	0.33	304.	281.00	278.00	290.60	5.15	108.91	C
20240.	3.64	0.47	304.	281.30	277.60	294.60	5.72	108.91	C
26630.	4.39	0.68	304.	281.50	277.10	300.00	6.34	108.91	C
8399.	2.08	0.06	408.	282.30	280.20	285.00	3.69	108.91	C
9523.	2.33	0.10	408.	282.50	280.20	286.70	4.03	108.91	C
12340.	2.85	0.16	408.	282.90	280.00	289.90	4.61	108.91	C
15980.	3.47	0.24	408.	283.20	279.70	292.60	5.15	108.91	C
20790.	4.23	0.35	408.	283.60	279.30	297.00	5.72	108.91	C
27290.	5.23	0.51	408.	283.90	278.70	302.70	6.34	108.91	C
10780.	1.91	0.01	196.	281.00	279.10	287.30	0.35	108.91	P
8502.	1.52	0.14	196.	280.60	279.10	285.30	0.97	108.91	P
20470.	4.66	0.07	303.	283.40	278.70	294.10	0.35	108.91	P
16180.	3.78	0.23	303.	282.30	278.50	290.30	0.97	108.91	P
12730.	3.04	0.34	303.	281.30	278.30	287.60	1.54	108.91	P
10000.	2.41	0.43	303.	280.40	278.00	285.20	2.06	108.91	P
17860.	4.35	0.13	389.	285.80	281.50	294.00	0.97	108.91	P
13810.	3.51	0.23	389.	284.80	281.30	291.00	1.54	108.91	P
10650.	2.81	0.30	389.	283.80	281.00	288.30	2.06	108.91	P
7655.	2.09	0.37	389.	282.70	280.60	286.50	2.66	108.91	P
6309.	1.72	0.40	389.	282.00	280.30	285.30	3.00	108.91	P
4580.	1.42	0.11	209.	279.30	277.90	282.10	4.03	108.91	C
5836.	1.59	0.18	208.	279.40	277.80	283.40	4.61	108.91	C
6954.	1.78	0.25	209.	279.50	277.80	284.50	5.15	108.91	C
8058.	2.00	0.33	209.	279.70	277.60	286.20	5.72	108.91	C
9210.	2.29	0.44	209.	279.80	277.50	288.00	6.34	108.91	C
5321.	1.56	0.11	210.	279.90	278.30	282.10	3.69	108.91	C

084.	1.68	0.15	210.	280.00	278.30	283.20	4.03	108.91	C
7628.	1.91	0.23	210.	280.10	278.20	285.00	4.61	108.91	C
9333.	2.17	0.32	210.	280.30	278.10	286.40	5.15	108.91	C
11370.	2.48	0.44	210.	280.50	278.00	288.90	5.72	108.91	C
13930.	2.87	0.60	210.	280.60	277.80	291.70	6.34	108.91	C
12050.	3.00	0.05	196.	280.60	277.60	288.00	0.35	108.91	P
9460.	2.54	0.19	196.	280.10	277.50	285.70	0.97	108.91	P
7423.	2.14	0.29	196.	279.60	277.40	284.00	1.54	108.91	P
5855.	1.81	0.37	196.	279.10	277.30	282.70	2.06	108.91	P
4410.	1.46	0.44	196.	278.60	277.20	281.60	2.66	108.91	P
5483.	1.62	0.11	210.	279.90	278.30	282.20	3.69	108.91	C
6397.	1.75	0.15	210.	280.00	278.30	283.40	4.03	108.91	C
8045.	2.00	0.24	210.	280.20	278.20	285.20	4.61	108.91	C
9674.	2.27	0.33	210.	280.30	278.10	286.70	5.15	108.91	C
11460.	2.59	0.46	210.	280.50	277.90	289.20	5.72	108.91	C
13530.	2.97	0.61	210.	280.70	277.70	292.00	6.34	108.91	C
10410.	1.85	0.13	204.	282.60	280.80	288.30	0.97	108.91	P
8164.	1.47	0.24	204.	282.10	280.70	286.50	1.54	108.91	P

APPENDIX B3

Convective Boiling of R1234ze within a micro-fin tube

(file: lowgwpq.dat)

q'' (Wm^{-2})	ΔT_s (K)	x_q	G_r (kg m^{-2}s^{-1})	T_w (K)	T_s (K)	T_f (K)	z (m)	M_w (g/mole)	flow
5061.	1.40	0.11	199.	280.60	279.20	283.10	4.03	114.04	C
6562.	1.56	0.18	199.	280.60	279.10	284.60	4.61	114.04	C
8383.	1.77	0.26	199.	280.70	278.90	286.00	5.15	114.04	C
10700.	2.03	0.37	199.	280.80	278.70	288.00	5.72	114.04	C
13750.	2.38	0.53	199.	280.90	278.50	290.60	6.34	114.04	C
6370.	1.89	0.06	305.	281.40	279.50	283.60	3.69	114.04	C
7141.	2.05	0.10	305.	281.50	279.50	284.90	4.03	114.04	C
9508.	2.46	0.16	305.	281.70	279.30	287.20	4.61	114.04	C
12910.	2.98	0.24	305.	281.90	278.90	289.40	5.15	114.04	C
17630.	3.67	0.36	305.	282.10	278.40	292.70	5.72	114.04	C
24260.	4.61	0.54	305.	282.30	277.70	297.40	6.34	114.04	C
7103.	2.04	0.03	410.	284.30	282.30	286.20	3.69	114.04	C
7744.	2.30	0.06	410.	284.50	282.20	287.60	4.03	114.04	C
10380.	2.91	0.11	410.	284.90	282.00	290.30	4.61	114.04	C
14580.	3.67	0.18	411.	285.30	281.60	292.90	5.15	114.04	C
20720.	4.68	0.29	410.	285.60	280.90	296.80	5.72	114.04	C
29560.	6.04	0.45	410.	286.00	280.00	302.80	6.34	114.04	C
10040.	2.95	0.01	206.	281.30	278.40	286.50	0.35	114.04	P
7868.	2.36	0.12	206.	280.60	278.30	284.60	0.97	114.04	P
6165.	1.87	0.20	206.	280.00	278.10	283.40	1.54	114.04	P
15050.	3.61	0.13	309.	285.60	282.00	291.70	0.97	114.04	P
11730.	2.79	0.24	309.	284.50	281.70	289.30	1.54	114.04	P
9077.	2.10	0.31	309.	283.50	281.40	287.10	2.06	114.04	P
12940.	4.25	0.01	305.	284.60	280.30	290.10	0.35	114.04	P
10140.	3.33	0.10	305.	283.50	280.20	287.80	0.97	114.04	P
7961.	2.56	0.17	305.	282.50	280.00	286.20	1.54	114.04	P
6297.	1.92	0.22	305.	281.60	279.70	284.70	2.06	114.04	P
12280.	2.07	0.40	213.	280.70	278.70	288.70	5.72	114.04	C
16120.	2.53	0.57	213.	280.90	278.30	291.80	6.34	114.04	C
6049.	1.96	0.08	297.	279.40	277.50	282.00	3.69	114.04	C
7177.	2.09	0.11	297.	279.50	277.40	283.40	4.03	114.04	C
9733.	2.48	0.18	297.	279.70	277.20	285.60	4.61	114.04	C
12830.	3.00	0.26	297.	279.80	276.80	287.80	5.15	114.04	C
16750.	3.71	0.38	297.	280.00	276.30	291.00	5.72	114.04	C
21910.	4.72	0.55	297.	280.10	275.40	295.30	6.34	114.04	C
6748.	2.14	0.05	411.	285.60	283.50	287.70	3.69	114.04	C
7610.	2.33	0.07	411.	285.80	283.50	289.20	4.03	114.04	C
10400.	2.83	0.12	411.	286.10	283.20	292.00	4.61	114.04	C
14510.	3.48	0.19	411.	286.30	282.90	294.70	5.15	114.04	C
20290.	4.34	0.30	411.	286.60	282.30	298.80	5.72	114.04	C
28450.	5.53	0.45	411.	286.90	281.40	304.90	6.34	114.04	C

11460.	3.24	0.01	210.	283.90	280.60	289.50	0.35	114.04	P
9093.	2.63	0.13	210.	283.20	280.50	287.40	0.97	114.04	P
7182.	2.12	0.22	210.	282.50	280.40	286.00	1.54	114.04	P
5656.	1.69	0.29	210.	281.90	280.20	284.60	2.06	114.04	P
14760.	4.00	0.10	313.	287.60	283.60	293.50	0.97	114.04	P
11370.	3.12	0.20	313.	286.50	283.40	291.20	1.54	114.04	P
8792.	2.40	0.27	313.	285.50	283.10	289.00	2.06	114.04	P
6460.	1.67	0.34	313.	284.30	282.60	287.40	2.66	114.04	P
10150.	3.31	0.09	300.	283.60	280.20	288.00	0.97	114.04	P
7983.	2.58	0.16	300.	282.60	280.00	286.40	1.54	114.04	P
6302.	1.99	0.21	300.	281.80	279.80	284.90	2.06	114.04	P
4638.	1.51	0.07	199.	279.10	277.60	280.90	3.69	114.04	C
5368.	1.63	0.11	199.	279.20	277.50	281.90	4.03	114.04	C
6950.	1.89	0.18	199.	279.30	277.40	283.50	4.61	114.04	C
8802.	2.19	0.27	199.	279.50	277.30	284.90	5.15	114.04	C
11100.	2.56	0.38	199.	279.60	277.10	287.20	5.72	114.04	C
14080.	3.04	0.54	199.	279.80	276.70	290.00	6.34	114.04	C
6196.	1.75	0.06	312.	280.90	279.20	283.00	3.69	114.04	C
6921.	1.90	0.09	312.	281.00	279.10	284.30	4.03	114.04	C
9062.	2.26	0.15	312.	281.20	278.90	286.50	4.61	114.04	C
12080.	2.75	0.23	312.	281.40	278.60	288.50	5.15	114.04	C
16250.	3.39	0.33	312.	281.50	278.10	291.80	5.72	114.04	C
22060.	4.29	0.49	312.	281.70	277.40	296.20	6.34	114.04	C
6817.	2.13	0.04	418.	283.80	281.70	285.80	3.69	114.04	C
7568.	2.38	0.06	418.	284.00	281.70	287.30	4.03	114.04	C
10180.	2.97	0.12	418.	284.40	281.40	290.00	4.61	114.04	C
14120.	3.71	0.18	418.	284.70	281.00	292.50	5.15	114.04	C
19730.	4.68	0.28	418.	285.10	280.40	296.50	5.72	114.04	C
27710.	6.01	0.43	418.	285.50	279.50	302.40	6.34	114.04	C
6920.	1.73	0.06	102.	279.60	277.80	284.30	0.35	114.04	P
5519.	1.44	0.21	102.	279.20	277.80	282.90	0.97	114.04	P
12640.	3.43	0.13	200.	282.80	279.40	288.90	0.97	114.04	P
9919.	2.73	0.27	200.	281.90	279.20	286.70	1.54	114.04	P
7747.	2.16	0.36	200.	281.10	278.90	284.70	2.06	114.04	P
5632.	1.56	0.44	200.	280.20	278.60	283.00	2.66	114.04	P
10380.	3.59	0.09	300.	282.20	278.60	287.00	0.97	114.04	P
8307.	2.83	0.17	300.	281.20	278.40	285.30	1.54	114.04	P
6636.	2.18	0.22	300.	280.30	278.20	283.70	2.06	114.04	P
4989.	1.50	0.27	300.	279.30	277.80	282.40	2.66	114.04	P
21120.	5.22	0.00	301.	287.10	281.90	296.80	0.35	114.04	P
16580.	4.27	0.16	301.	285.90	281.60	292.80	0.97	114.04	P
12870.	3.44	0.28	301.	284.80	281.40	290.00	1.54	114.04	P
9862.	2.74	0.36	301.	283.80	281.10	287.50	2.06	114.04	P
6879.	2.00	0.43	301.	282.60	280.60	285.70	2.66	114.04	P
5456.	1.61	0.46	301.	282.00	280.40	284.60	3.00	114.04	P
8354.	1.73	0.26	206.	280.30	278.50	285.70	5.15	114.04	C
10380.	2.02	0.36	206.	280.40	278.30	287.80	5.72	114.04	C
12970.	2.39	0.51	206.	280.50	278.10	290.40	6.34	114.04	C
6403.	1.94	0.06	302.	281.10	279.10	283.30	3.69	114.04	C

7228.	2.12	0.09	302.	281.20	279.10	284.70	4.03	114.04	C
9500.	2.55	0.16	302.	281.40	278.90	287.10	4.61	114.04	C
12600.	3.09	0.24	302.	281.70	278.60	289.20	5.15	114.04	C
16790.	3.81	0.36	302.	281.90	278.10	292.60	5.72	114.04	C
22580.	4.78	0.52	302.	282.10	277.40	297.20	6.34	114.04	C
6940.	2.21	0.05	404.	284.10	281.90	286.20	3.69	114.04	C
7755.	2.47	0.07	404.	284.30	281.80	287.70	4.03	114.04	C
10360.	3.07	0.13	404.	284.70	281.60	290.50	4.61	114.04	C
14150.	3.81	0.20	404.	285.00	281.20	293.10	5.15	114.04	C
19480.	4.79	0.30	404.	285.40	280.60	297.00	5.72	114.04	C
26980.	6.11	0.45	404.	285.80	279.60	302.80	6.34	114.04	C
6499.	1.65	0.05	103.	279.20	277.50	283.60	0.35	114.04	P
11870.	2.92	0.00	203.	282.30	279.40	288.70	0.35	114.04	P
9673.	2.35	0.14	203.	281.70	279.30	286.50	0.97	114.04	P
7790.	1.88	0.24	203.	281.00	279.20	284.90	1.54	114.04	P
6179.	1.49	0.31	203.	280.50	279.00	283.30	2.06	114.04	P
12490.	3.05	0.17	208.	283.70	280.60	289.60	0.97	114.04	P
9769.	2.55	0.30	208.	283.00	280.50	287.40	1.54	114.04	P
7474.	2.12	0.39	208.	282.40	280.30	285.40	2.06	114.04	P
5066.	1.67	0.46	208.	281.70	280.00	284.10	2.66	114.04	P
3848.	1.44	0.49	208.	281.20	279.80	283.30	3.00	114.04	P
10020.	3.49	0.07	296.	282.20	278.70	286.70	0.97	114.04	P
8076.	2.71	0.14	296.	281.20	278.50	285.00	1.54	114.04	P
6464.	2.07	0.19	296.	280.30	278.20	283.50	2.06	114.04	P
19900.	4.64	0.04	300.	286.70	282.10	296.40	0.35	114.04	P
15780.	3.71	0.19	300.	285.60	281.90	292.60	0.97	114.04	P
12380.	2.91	0.30	300.	284.50	281.60	289.90	1.54	114.04	P
9607.	2.24	0.38	300.	283.50	281.30	287.50	2.06	114.04	P
7115.	1.68	0.02	100.	281.90	280.20	286.40	0.35	114.04	P
10590.	3.00	0.02	203.	280.90	277.90	286.60	0.35	114.04	P
8544.	2.40	0.14	203.	280.20	277.80	284.50	0.97	114.04	P
6840.	1.90	0.23	203.	279.60	277.60	283.10	1.54	114.04	P
5435.	1.50	0.29	203.	279.00	277.50	281.80	2.06	114.04	P
13940.	3.62	0.05	204.	282.30	278.70	290.00	0.35	114.04	P
11250.	2.98	0.20	204.	281.60	278.60	287.40	0.97	114.04	P
8941.	2.45	0.32	204.	280.90	278.40	285.40	1.54	114.04	P
6974.	2.01	0.40	204.	280.20	278.20	283.60	2.06	114.04	P
4888.	1.56	0.47	204.	279.50	277.90	282.30	2.66	114.04	P

www.ingramcontent.com/pod-product-compliance
Lightning Source LLC
Chambersburg PA
CBHW081858170526

45167CB00007B/3062